THE NEW
CHALLENGE
OF THE STARS

PATRICK MOORE AND DAVID HARDY
FOREWORD BY ARTHUR C. CLARKE

The 'ring of fire' *frontispiece*
The most essential requirement for the inhabitants of the planet Earth, for it encompasses all life on our world. It is our atmosphere, which refracts sunlight during an eclipse of the Sun, seen from a vantage point some thirty miles above the surface of the Moon.

Rand McNally & Company
New York Chicago San Francisco

Contents

Editor
Lawrence Clarke

Art Editor
John Ridgeway

Production
Julian Deeming

The Rand McNally New Challenge of the Stars
Edited and designed by
Mitchell Beazley Publishers Limited
© Mitchell Beazley Publishers Limited 1977
© David A. Hardy 1972 and 1977
39 original illustrations by David A. Hardy
All rights reserved
ISBN 528 83087 2
Library of Congress Catalog Card Number 71 18909
Published 1978 in the United States of America by
Rand McNally & Company, P.O. Box 7600,
Chicago, Illinois 60680

Typesetting by Tradespools Limited, Frome
Reproduction by Gilchrist Brothers Limited, Leeds
Printed and bound in Great Britain by
Hazell Watson and Viney Limited

Foreword

Everyone knows the role that science-fiction writers have played in presenting the idea of space travel to the world decades — indeed, centuries — before it was actually achieved. It might well be argued that if there had been no writers, there would be no astronauts today. Before the reality, there must be the dream to provide inspiration.

The role of the space artists, however, has been much less appreciated, perhaps because there are so few of them; the good ones can be counted on the fingers. I do not know who first tried to visualize the Earth as a planet; it could well have been the ubiquitous Leonardo, whose notebooks contain some remarkable 'aerial' views of cities and landscapes. It would seem unlike him, not to have taken these to their logical conclusion.

Popular books on astronomy, over the last few centuries, must occasionally have regaled their readers with space-views of the Earth and Moon. Perhaps the most outstanding example is Nasmyth and Carpenter's classic *The Moon* (1874) which employed a combination of plaster models and photography to give strikingly realistic views of lunar landscapes. However, it was the rise of the science-fiction magazines in the 1930s that first gave the space-artist a popular audience; scanning back through my memories, the very first science-fiction magazine I ever saw was a *circa* 1929 issue of *Amazing Stories*, with a cover showing Jupiter hanging above the landscape of one of its satellites. The artist was Frank R. Paul — whose exotic space-vistas, bug-eyed extraterrestrials, and apparently cloned humans will be remembered with affection by all old-timers.

It must be admitted that most of the pulp-magazine artists — who had little knowledge of science, and were probably even more miserably paid than the writers — were concerned with entertainment rather than with accuracy. That both could be combined was demonstrated when *Life* Magazine, in the early 1940s, published Chesley Bonestell's stunning views of Saturn from Titan, Mimas and its other moons. I can still recall the impact of those paintings — and my annoyance because some earth-bound *Life* editor had remarked of the tiny figures in Chesley's moonscapes that they were merely put in 'to give scale'. To give scale, indeed! Didn't he have the imagination to realize that some day men would actually *be* out there, looking up at the ringed glory of the most spectacular of all the planets?

Though it will be some little while yet before human explorers reach Saturn, we are now in an interesting transition period when we can compare the realities of space with the earlier imaginings of the artists. The Gemini photos of Earth, the Orbiter and Apollo photos of the Moon — these have largely confirmed, but have not superseded, the creations of the astro-painters. The camera (despite many nineteenth century fears) failed to displace the easel on *this* planet; nor will it do so in space. In fact, we have already had one orbiting painter (Cosmonaut Alexei Leonov), and there will be many others in the years to come.

But the astronomical artist, like the writer, will always be far ahead of the explorer. He can depict scenes which no human eye will ever witness, because of their danger, or their remoteness in time or space. Only through the eye of imagination can we watch the formation of the planets, the explosion of a super-nova, the ball-bearing smooth surface of a neutron star, or the view of our own Galaxy, looking back from its off-shore islands, the Clouds of Magellan.

Of course, the space-artist can sometimes be proven wrong — but that is part of the fun. Pre-Apollo moonscapes were invariably too jagged, and few if any artists (or scientists) anticipated the softly-yielding soil of the *maria*. Nor did anyone expect Mars to be covered with craters that would have looked perfectly at home on the Moon; or a Venus that rotated backwards and was almost red-hot; or a Mercury that had sunrise and sunset. There will be other surprises to come.

I welcome this book by David Hardy and my old friend Patrick Moore, because it provides a heady mixture of education, entertainment — and inspiration. We need the latter, now that the excitement of the Apollo program has ebbed, and so many voices are asking 'Why go to the planets?' David Hardy's poetically luminous paintings are a reminder that the Moon is not the goal, but merely the beginning, of *real* space exploration.

This book also gives me a reassuring feeling of progress — of the continuity of human effort. It is now almost twenty years since I was working, with the late R. A. Smith, on the text and illustrations of *The Exploration of the Moon*, a forecast of the early days of manned spaceflight. That book has long been out of print, because most of it is now past history; but it is sometimes fun to glance back over its pages.

In the same way, I look forward to checking the accuracy of David Hardy's visions — twenty years from now.

Arthur C. Clarke
New York

Introduction: Fiction or Fact?

When Neil Armstrong stepped out onto the rocks of the Moon, on July 16, 1969, he ushered in a new era — the Age of Space; an age that would have seemed sheer fantasy only a few decades ago.

As fiction turns slowly but inexorably into fact, it is hard to tell where the dividing line can be drawn — if indeed it exists at all. Can we entirely discount the idea of murderous aliens described so vividly by H. G. Wells in his classic *The War of the Worlds*, or in the interstellar battles in more recent epics such as *Star Wars*? The mind that created the former foresaw a technology that eventually placed Man on the Moon; perhaps some of the technology envisaged in *Star Wars* will not appear so fantastic in years to come.

'Invasion from space' is a favorite theme in science fiction; H. G. Wells is by no means the only writer to have used it. Fictional aliens are always technologically advanced, and generally hostile. This has been particularly true of Martians, from the planet associated with the 'God of War.' In reality observational studies suggest that at least one-third of all stars have planetary companions that could harbor such aliens. It may not be so extraordinary to consider what our reactions would be if the Earth were visited from beyond the Solar System.

Undoubtedly there would be both suspicion and fear. We cannot yet send probes to the planets of other stars, and we would have to realize at once that we were being faced with beings far in advance of ourselves from a purely technological point of view. It would be only too easy to act in a way that would make the aliens regard us as unfriendly, with potentially dangerous results.

As yet there is no undisputed evidence that the Earth has been visited. However, it seems highly probable that intelligent life is scattered widely throughout the universe. The Sun is a

run-of-the-mill star, and is one of 100,000 million in our Galaxy alone; all the evidence suggests that planetary systems are common, and it would be sheer conceit to assume that ours is the only civilization. If a planet similar to the Earth moves round a star similar to the Sun, why should not life evolve there on the same pattern as our own?

Although alternative life forms conceived in fiction are infinite, the evidence, so far as it goes, indicates that totally alien forms are improbable. If there are beings elsewhere that can, for instance, survive in the total absence of atmo-

sphere, then most of our modern science is wrong. However, there is no reason why an extraterrestrial being should look like a man or a woman. There is no reason why an intelligent astronomer living on some faraway world should not have three legs and four arms. Everything depends upon the environment; for example, if the force of gravity at the planet's surface differs from that of Earth, then life will develop differently.

The fact that there is no other advanced life in our Solar System is due solely to the fact that the other planets are not suitable for it. When we look

farther afield, we are confronted with the problem of sheer distance.

Efforts have already been made to pick up radio transmissions from other worlds. They have not been successful, but they may one day lead to positive results. But before that day man will have to heed Wells's advice in the implied analogy — the Martian invaders treated man as contemptuously as man treats his own minority groups or now endangered animal species. Perhaps making contact with aliens is unwise; however, if humanity persists, we cannot expect to maintain our isolation.

War of the Worlds *above*
'A monstrous tripod, higher than many houses, striding over the young pine-trees, and smashing them aside in its career; a walking engine of glittering metal, striding now across the heather; articulate ropes of steel dangling from it, and the clattering tumult of its passage mingling with the riot of the thunder. . . . Vast spider-like machines, capable of the speed of an express train, and able to send out a beam of intense heat.' Such were the Martian fighting machines described by H. G. Wells.

The scene of the ironclad *Thunder Child* making a last suicidal bid to ram an invader is pure science fiction. In the Solar System there can be no 'war of the worlds'. But who can tell what alien races, and what conflicts, exist elsewhere in our Galaxy?

Space Stations

The twentieth century has been the Age of Challenge. In its earliest years, Man took to the air; since then he has explored the ocean bed, and has reached the summits of our highest mountains. In the last decade, he has done more. He has reached the Moon, and has sent his messengers out to other planets. We have lived through the beginning of the greatest challenge of all: the invasion of outer space.

Much has been achieved already, and critics of the space program conveniently forget the immense practical benefits which have become apparent. Communications satellites have become part of our everyday life; many lives have been saved by the weather satellites, which have given advance warnings of dangerous tropical storms; all branches of science have shared in the profits. Yet this is only a beginning. In this book we are looking to the future – to the time when we reach not only the planets, but also the stars. It is a breathtaking prospect.

The idea of space-travel is not new. Indeed, it goes back to at least the second century A.D., when a Greek satirist, Lucian, wrote a story about a voyage to the Moon. Other writers followed in Lucian's footsteps, but it was not until our own time that space-travel became a practical possibility. Konstantin Tsiolkovskii, a shy, deaf Russian teacher, wrote about it in scientific vein just about the time that the Wright Brothers were making their first 'hops' at Kitty Hawk; and yet after the First World War Robert Goddard, the American who fired the first liquid-propellant rocket in history, was ridiculed for daring to suggest that small vehicles might be sent as far as the Moon.

At that time, of course, rockets were both feeble and unreliable. There seemed no other method of reaching space; the atmosphere of the Earth extends upward for a very limited distance, and no ordinary flying machine will function in a vacuum. This limitation does not apply to a rocket, which works according to the principle of reaction: every action has an equal and opposite reaction. As gas is sent out through the rocket exhaust, so the vehicle itself is propelled in the opposite direction. The principle is exactly the same as that of the firework display rocket, though for space-probes a charge of gunpowder is replaced by a vast, immensely complicated rocket motor powered, in the main, by liquid propellants.

Rocket research
Rocket research was undertaken in Germany in the decade before the Second World War by men such as Willy Ley and Wernher von Braun, and the first really effective high-altitude rockets, the V2 weapons, came from the Peenemünde station in the Baltic. After the war, many of the German team went to the United States to continue their scientific research. The captured V2's were put to good use. In 1949 a compound arrangement known as a two-stage rocket was launched from the proving ground at White Sands. It consisted of a V2 carrying a smaller rocket, a WAC Corporal, which was given a 'running jump' into space and reached the record height of over 244 miles.

Meanwhile, rocket work was going on in the Soviet Union. On October 4, 1957 the Russians achieved a startling success; they launched Sputnik I, the first man-made moon or artificial satellite in all history. It was a tiny thing, only about the size of a football, but it marked the opening of the Space Age. As it sped round the world, sending back its 'Bleep! bleep!' signals, nobody could doubt that the Moon was within reach.

Satellites of many kinds followed Sputnik I. Up to now most of them have been Russian or American, though small satellites have also been launched by Britain, Japan, China and France. The satellites have contributed immensely to our knowledge of the Earth itself and of the universe. The detection of the all-important Van Allen zones of intense radiation around the Earth is only one of the many discoveries which the satellites have made possible.

Men in space
Just as the Russians had launched the first artificial satellite, so they sent up the first man into space: Yuri Gagarin, who made his flight on April 12, 1961. Though he completed only one circuit of the Earth, his journey silenced the critics who had claimed that an astronaut would be at once affected by what fiction-writers commonly called space-sickness, due to the condition of zero gravity. Gagarin was followed by other space-travellers from the U.S.S.R. and the U.S.A.; and it is significant that the first American venture into space was made by Alan Shepard, who, less than ten years later, set foot upon the Moon.

As the 1960s passed by, the original single-passenger space-vehicles were replaced by larger and more complicated craft, carrying two men (as with the Gemini vehicles) or even three (as with the Apollo vehicles and some of the Russian space-ships). Docking maneuvers were successfully carried out, as were what are usually called 'space-walks'. Of course there were tragedies. Three American astronauts – Grissom, White and Chaffee – were killed at Cape Kennedy after a disastrous fire in a capsule which was being tested on the ground; and not many weeks later Vladimir Komarov, a veteran Russian cosmonaut, crashed to his death when returning from a spaceflight. Then, in 1971, came the tragic deaths of Cosmonauts Dobrovolsky, Volkov and Patsayev, who had spent over three weeks in the massive space-station Salyut; their capsule landed gently, but the three passengers were dead. Their fate was a grim reminder that space is an alien environment, and that the risks of entering it are great indeed.

Laboratories in space
Salyut was the first true space-laboratory. Since then the Americans have launched Skylab, an ambitious research base orbiting the Earth and manned by relays of crews, each crew remaining in the base for a number of weeks. Sky-lab is made up essentially of hardware of the type used in the Apollo lunar mission. The orbital workshop is a modified Saturn 4b stage, attached to which are solar panels to provide the power, an airlock module and multiple docking adaptor, and an Apollo Telescope Mount.

The main Skylab program ended on February 8, 1974, when the members of the third and final crew — Astronauts Carr, Gibson and Pogue — splashed down in the Pacific after having been in space for a record period of over eighty days. Much had been accomplished.

Earth resources, for instance, were studied by means of six cameras, each of which recorded data from the same areas of the Earth's surface but operated on different wavelengths; to give just one example, regions of diseased crops could be identified at once. Information was gathered with regard to mineral deposits, water supplies, seasonal variations in temperature, snow cover, potential flood areas and much else.

Astronomy benefited considerably, and careful attention was paid to the Sun. Experiments using the Apollo Telescope Mount gave basic information which had been previously unobtainable. The first mission alone revealed more information about the outer corona than all the hours of observation of the phenomenon at natural eclipse in over one thousand years. The continuous record over a total period of 257 days also gave a unique account of the variation of the daily relative sunspot number – used as a measure of solar activity. Solar flares were measured, and the X-radiations were studied more completely than ever before. Other bodies were also studied, in particular Kohoutek's Comet, which had been disappointing seen from Earth, but which was found from Skylab to be surrounded by a vast cloud of tenuous hydrogen.

Even engineering and manufacturing experiments were conducted; welding and casting of metals was undertaken, and the zero gravity conditions led on to new discoveries. For instance, it was thought likely that perfect crystals might form under weightless conditions.

Projects for the future
Up to recent times one of the main difficulties in the way of intensive space research has been the sheer cost. In America the space budget was drastically cut after the end of the Apollo program, and there are grounds for supposing that the same has been true of the Soviet program. However, there have been encouraging signs – notably the 1975 space docking between an American and a Russian craft — which have paved the way for future collaboration in space.

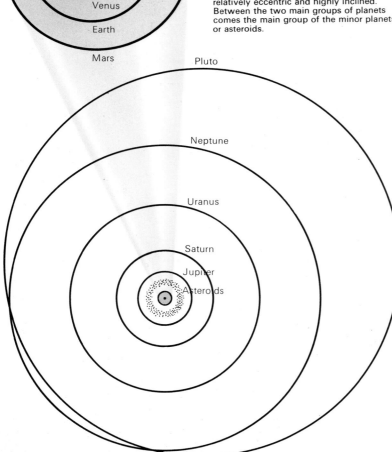

The planets of our Solar System *left*
The inner planets: Mercury, Venus, the Earth and Mars. *Below* The outer planets: Jupiter, Saturn, Uranus, Neptune and Pluto. The distances from the Sun range from an average of 36,000,000 miles for Mercury out to over 3,500,000,000 miles for Pluto. It is clear that the System is divided into two distinct parts; the inner planets are relatively small, and are essentially of the same type as the Earth, while the four giants are gaseous, at least in their outer layers. Pluto is something of a mystery; it is apparently no larger than Mars, and has an exceptional orbit, relatively eccentric and highly inclined. Between the two main groups of planets comes the main group of the minor planets or asteroids.

Disaster averted *right*
Skylab, launched on May 14, 1973, was a true space-laboratory. Its first mission was by no means uneventful. Just over a minute after launch, a malfunction caused the loss of the thin aluminum meteoroid shield; in orbit, one of the solar panels became entangled, while the other was found to have been ripped off during the launch maneuver. On May 25 three astronauts made a rendezvous with Skylab and began emergency repairs, finally completed by the second crew in June.

Space Cities

Just as the 1950s, 1960s and 1970s saw the steady development of rocket probes and artificial satellites, so the 1980s will see the development of massive orbital space-stations. They represent the next phase of Man's advance beyond the Earth, and by now detailed plans have been drawn up. Russia's Salyut, and the U.S. Skylab, may be regarded as pioneer attempts; and despite the tragedy of Soyuz 11, when the first Salyut crew met with disaster after spending more than three weeks on the space-station, the outlook is encouraging for permanently orbiting laboratories and observatories.

The space-stations will be used primarily as research bases, and all branches of science will benefit immeasurably. Yet there is another aspect to be borne in mind. Future orbital stations will be used as starting-bases for interplanetary flights to Mars, Venus and beyond. We are no longer bound to our home planet.

When scientific rocket research began, as long ago as the years before the last war, it was both expensive and wasteful. A rocket could be used only once. It was launched; it spent a very brief period in the upper atmosphere, and then it fell back to the ground, destroying itself and often also destroying the equipment which it carried. This limitation persisted with the V2's which were taken to America after 1945, and also with other scientific rockets such as the Vikings. What was needed, of course, was a vehicle which would stay aloft for a protracted period, sending back its findings; and this was achieved with the artificial satellites launched from 1957 onward.

At that time the so-called 're-entry problem' was regarded as one of the most difficult in astronautics. If a spacecraft comes back into the Earth's atmosphere too quickly, it will burn away in the manner of a meteor. The velocity of re-entry must be exactly right. In the event, the problem was solved surprisingly quickly; but it was still true to say that undue waste was involved, since no vehicle could make more than a single flight into space. Clearly, the next step was to produce a vehicle which could be used over and over again.

The space shuttle
The Shuttle has been described as a cross between an aircraft, a spaceship and a glider. This is a recoverable vehicle, which can be used time and time again, and will act as a ferry be-tween Earth and orbiting space stations. Preliminary tests, carried out in 1977, held high hopes for the future. The Shuttle also means that astronauts will not necessarily have to be so highly 'space trained;' the way will be open for scientists of all disciplines.

Unless there are any unforeseen complications, the American shuttle should be operational by 1980. The vehicles could have many roles to play. For instance, they will be able to carry up relatively large satellites and put them into orbit. They will also be used for inspecting and repairing others.

However, the main function of the shuttles will be to deliver propellant modules for lunar or planetary missions and, above all, to ferry crews to the orbital stations and back home. In the painting, one shuttle has delivered a propellant module for a manned nuclear space-craft, one of two preparing for a voyage to Mars; the nearer shuttle is firing its retro-rockets for return to Earth.

The space station
The station shown here, based on a recent McDonnell-Douglas study, is a development of the Skylab. It is assembled, in a 270-nautical mile orbit, from modules launched by powerful two-stage vehicles. It will accommodate 50, then up to 100 technicians and scientists of all disciplines. Because of the potential hazards of long periods under zero gravity, the station is designed to rotate. Zero gravity conditions will prevail at the hub, but 'artificial gravity' will be provided on other parts of the station; it will spin once every 15 seconds at a radius of 100 feet, but large portions will be contra-rotated to facilitate docking, and for scientific experiments which have to be carried out under conditions of complete weightlessness. Other modules, with astronomical telescopes and other instruments – shown to the right – are free-flying because they require very accurate pointing or very low gravity levels. The main station is, of course, nuclear-powered, and contains a closed life-support system.

To list all the uses of the station would take many pages. One, of practical importance, is brought out in the painting. As we can detect, the clouds show the pattern of a depression. Before the Space Age, meteorology was a delightfully uncertain science, because all observations had to be made from below the atmosphere; it was impossible to gain anything like an overall picture. From space, however, whole weather systems can be studied, and their behavior interpreted. Many lives have already been saved in this way; dangerous tropical storms far out at sea have been photographed from weather satellites, and people who live in the affected coastal zones have been given ample warning. From the manned space-station a continuous 'storm patrol' will be maintained.

Earth's atmosphere contains protective layers which screen off many of the radiations which bombard us from space. From our point of view this is just as well, as some of them are dangerous; but the screening imposes severe limitations on both astronomers and physicists. From the orbital station they will be able to study the whole range of frequencies. Neither must medical research be neglected; indeed, this may prove to be one of the most important aspects of the entire program. It is even possible that surgical operations impossible to perform on Earth will be carried out under the zero-gravity environment of the space-station.

The space city
Skylab was the first genuinely scientific space station, and it will be succeeded by Spacelab, which will be more ambitious still. But in the foreseeable future there may well be veritable 'space cities' – a prospect suggested long ago by the Russian Konstantin Tsiolkovskii and since elaborated upon by the American physicist Gerard O'Neill and others.

The space cities are envisaged as housing more than 1,000 people in a nuclear-powered closed-cycle system, in which there is neither leakage nor loss. As scientific bases they will be invaluable; astronomy, for instance, will benefit because of the lack of surrounding atmosphere, but there will be practical advantages too. Transworld communication problems will be solved, and there is also the exciting possibility of beaming solar energy onto the Earth by means of microwave techniques, so that solar power from the orbital cities may replace the wasteful expenditure of fossil fuels — and also the potentially dangerous widespread use of nuclear power stations. From the purely commercial point of view, the cities will have the advantage of 'free' vacuum, limitless solar power and wide temperature ranges, which will enable experiments to be carried out under conditions which cannot be reproduced on Earth.

There are, however, two aspects which must be considered. First, there are the possibilities of using space cities for military purposes. We may hope that before elaborate orbital bases are set up the fear of global war will have receded; although it would appear that the military role of a space city could be potentially great. Still much less under our control is the question of how the human body will react to conditions in space. Screening the harmful radiations must be carried out, particularly at times when the Sun is particularly active and both radiations and high-velocity particles are sent out from the violent 'solar storms'. Solar maxima occur about every 11 years; the next is due near 1980.

If all goes well, then before the end of our century there will be many massive space-stations orbiting the world; in our skies they will become as familiar as the Sun, the Moon and the stars. They, above all, will represent the triumph of Man's ingenuity, and his ability to conquer all obstacles which stand in the way of scientific progress.

The space shuttle *below*
The original prototype for the shuttle spacecraft incorporated expendable booster rockets. In the future the rockets are likely to be replaced by a booster vehicle that can be reused in the same way as the orbiter.

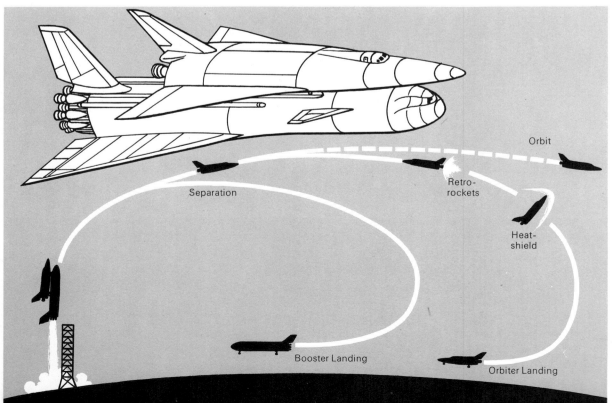

Separation

Orbit

Retro-rockets

Heat-shield

Booster Landing

Orbiter Landing

An orbiting space-station *right*
The station moves in an almost circular path at a height of 270 nautical miles above ground level; it can accommodate up to 100 crew-members. Below, we see a shuttle which has delivered a propellant module for a space-ship bound for Mars. The nearer shuttle is just firing its retro-rockets for return to Earth. Both shuttles will be re-usable, perhaps for hundreds of journeys into space; they are delta-winged, since, unlike 'pure' space-craft, they have to be designed so as to fly and maneuver in the Earth's atmosphere as well as in a vacuum. These shuttles are based on recent design studies by North American Rockwell with the British Aircarft Corporation.

The First Lunar Base

On July 21, 1969, the first men from Earth stepped out on to the barren surface of the Moon. Apollo 11 marked the greatest triumph of the program which had been initiated so many years earlier; it showed that man could not only reach the Moon, but could survive there.

The Moon is not a very friendly world. Airless, waterless and lifeless, its unprotected surface offers little comfort to colonists from Earth. Yet it represents an ideal site for a research base, and a step on to the planets. Our exploration of deep space can only begin with the Moon.

The Moon is a smaller world than the Earth. Its diameter is only 2,160 miles, and it has a mass of only 1/81 of that of our world. This means that it has a lower escape velocity (1.5 miles per second) and a lower surface gravity; an astronaut on the Moon has only one-sixth of his Earth weight. Walking on the Moon is easy enough, as has been amply demonstrated by the Apollo astronauts, but prolonged exertion is tiring.

During the 1960s the first really accurate maps of the Moon were drawn, from the photographs sent back by the Orbiter probes. These maps covered both the familiar hemisphere of the Moon and also the far hemisphere, which is never visible from Earth because it is always turned away from us. Then came the pioneer landings, and the first automatic lunar transmitting stations.

Fears had been expressed that the lunar surface might be too soft and dusty to be safe for manned exploration. These fears have not been justified; the rocks of the Moon are pleasingly firm, and there is no reason to suppose that treacherous regions exist.

Originally it was thought that because of the Moon's lack of atmosphere, meteoric bombardment might make it necessary to build a base underground, but this does not now seem to be the

The Moon as seen from the Earth. *right*
The broad dark plains are known as 'seas', even though they are waterless; the Moon is a world of lava-plains, mountains, peaks and craters of all sizes.

case. Probably the pioneer base will look rather like a group of cylinders or Antarctic survey huts; at first the living accommodation will be inside a grounded space-station which had formerly been in orbit round the Moon. This sort of base may well be established within the next ten years. It will not be permanently manned, but will be able to support crews who will stay in it for a period of weeks or even months at a time. Again, it is to be hoped that the base will be international.

The advanced base

From these humble beginnings, Man will plan a much more elaborate lunar colony; and our next picture shows a scene on the Moon during the twenty-first century. We are looking down several thousand feet into a small crater near the Moon's North Pole, where a permanent base has been set up. The larger dome is multi-storied, and contains the living quarters; above it is an optical telescope, and around it are the transparent tubes which make up the lunar 'farm'. Hydroponic methods will be very important on the Moon; the plants are nourished by liquids circulating beneath them. Even with the improved methods of transport between the Earth and the Moon, every effort must be made to make the lunar colony as self-supporting as possible, even though the Moon has a depressing lack of useful materials; in particular there are no hydrated substances. The remaining domes contain laboratories, equipment of all kinds, and stores. In the foreground, a 'crawler' vehicle is collecting samples from the inner slopes of the crater.

Though the Moon is so different from the Earth in its condition, it is made of essentially the same materials; the rocks are of the basaltic variety, and no unknown substances have been found there. Whether there will be any materials of sufficient commercial value to be ferried home seems doubtful. But to the geologist, the Moon is an ideal site for research. Since the lunar atmosphere was lost many thousands of millions of years ago, there has been virtually no erosion, as it occurs on Earth; the Moon's surface looks much the same now as it did before the time when the Earth was ruled by giant dinosaurs.

Though the Moon's surface does show slight tremors, powerful 'moon-quakes' belong to the remote past, and there is no danger that the lunar bases will be shaken down. Neither do meteoritic falls appear potentially very dangerous, even though the Moon has no atmosphere to act as a screen. On the other hand, the surface is exposed to all the various radiations coming from space; and protective methods to counter radiation effects will have to be devised.

Of course, the main disadvantage of the Moon is its lack of air, and the fear of a leak in the base will be ever-present in the colonists' minds. However, there will be no difficulties in communicating with the Earth, and there will be many artificial lunar satellites to act as radio and television relays between the different bases on the Moon itself.

Research activity

Life in the lunar colony will be full of interest. The Earth, looming large in the lunar sky, seems very far away; it will seem to stay almost motionless in the heavens, with the background of brilliant, untwinkling stars sweeping slowly past it. Astronomers living on the Moon will have great advantages; optical telescopes will not be handicapped by having to look through a layer of unsteady, obscuring atmosphere, and neither will radio telescopes be limited. Moreover, the lower surface gravity means that in some ways it will be easier to build large pieces of equipment, and this will be of particular use with radio telescopes, which have to be of considerable size. On the other hand there are great ranges of temperature on the Moon – from above +210 degrees Fahrenheit at noon down to −250 degrees Fahrenheit at midnight on the lunar equator, though near the Moon's pole the daytime heat will be much less. Great care must be taken in the choice of materials which will be exposed to these changes in temperature. But these difficulties will be overcome; the permanent lunar base will become reality, perhaps in our own time.

The pioneer lunar base *left*
This may well be set up within the next couple of decades. In the foreground we see the basic space-station module, placed on the lunar surface by the propulsion module of a 'space tug' similar to the one seen landing on the right. The grounded station can accommodate a crew of twelve, and can provide everything necessary for a prolonged stay. Behind is a cargo-landing craft, together with a separate cargo module. At the extreme right is a lunar drill, and in the right foreground is a 'Moon Rover' for long-range exploration. The bright 'star' to the left of the crescent Earth is Venus.

A permanent lunar base *right*
Near the Moon's North Pole, where the daytime temperature is much lower than at sites closer to the equator – though the nights are equally cold. In addition to the main dome, containing the living quarters, there are various others, used for equipment and storage. Each dome has its separate system of air-locks. In this scene the Earth has just passed through the Milky Way into the constellation of Gemini, the Twins. The red star to the right is the semi-regular variable star Eta Geminorum, shown at its maximum of just below the third magnitude.

Mars: After Viking

Beyond the Moon we come to the red planet, Mars — perhaps the most fascinating of all the worlds in the Sun's family. Though it is slightly more distant than Venus, it has always been regarded as our second space target, because it is more Earth-like than any other planet.

Less than a century ago it was widely believed that intelligent life might exist upon Mars. Percival Lowell, founder of the great observatory at Flagstaff in Arizona, was firmly convinced that the so-called canals that he observed on the red Martian deserts were artificial waterways. And later, it was regarded as highly probable that the dark areas, clearly visible from Earth, were tracts of vegetation.

Our view today is different. There are no Martians and no canals; neither are there extensive areas of vegetation. Yet in many ways Mars is not overwhelmingly hostile, and it will indeed be strange if it has not been reached within the next hundred years.

Mars and the Earth compared. *above* Earth has a diameter of 7,926 miles; Mars, only 4,200 miles. The escape velocity is much lower (3.1 miles per second), and the mass only one-tenth of that of the Earth.

When Mars is at its brightest, it shines more brilliantly than any other planet apart from Venus. It looks almost like a tiny red lamp in the sky, and it is not surprising that in earlier years attempts were made to signal to the 'Martians.' Only since the start of the Space Age have we been able to draw up a more reliable picture of what this intriguing world is really like — and astronomers have had many surprises.

With its diameter of 4,200 miles and its mass of only one-tenth that of the Earth, Mars is one of the smallest planets in the Solar System; only Mercury and Pluto are inferior to it. The escape velocity is only 3.1 miles per second, which is insufficient to retain a dense atmosphere similar to our own. Yet we have now found strong evidence that running water existed there only a few tens of thousands of years ago, in which case the atmosphere must have been much thicker than it is at present.

Telescopically, Mars shows a reddish disk, with permanent dark patches here and there, and white caps covering the poles. Maps were drawn up more than a century ago, and the various features were given romantic names; among these are the Mare Sirenum, or 'Sea of the Sirens,' and the Margaritifer Sinus, or 'Gulf of Pearls'. In the light of modern space-probe research this nomenclature has had to be revised. In particular, there are no seas on Mars. Liquid water cannot exist under so low an atmospheric pressure.

The Martian year amounts to 687 Earth days, and the tilt of the axis is almost the same as ours (between 23 and 24 degrees), so that the seasons are of the same general type, though they are naturally much longer. The rotation period is 24 hours 37 minutes. A Martian day has now become known as a 'sol,' making 669 sols in the Martian year.

Exploratory probes

Before the first successful probes, it was assumed that Mars had a gently undulating surface lacking in major mountains or deep valleys; it was also thought that the polar caps, which vary according to the Martian seasons, were due to a very thin layer of ice or frost only a millimeter or two in depth. The first results from Mariner 4, in 1965, caused a quick change of opinion. Mars proved to be crater-scarred, and it was at once evident that the dark regions were not due to vegetation; neither were they depressed — indeed the most prominent of them, the V-shaped Syrtis Major, is lofty. Even more important were the

measurements sent back with regard to the atmospheric density. The ground pressure was found to be unexpectedly low, and it followed that most of the atmosphere was made of up of carbon dioxide rather than nitrogen.

Mariner 4 was a flyby, and made only one active pass of Mars before entering a permanent orbit round the Sun. Mariners 6 and 7 followed in 1969, confirming the earlier findings. Then, in 1971, came the flight of Mariner 9, which was put into orbit round Mars and sent back thousands of high-quality pictures. For the first time the giant volcanoes and the yawning canyons were shown.

Surprise followed surprise. In the region known as Tharsis there were volcanoes which surpassed anything on the Earth; the highest of them, Olympus Mons (formerly known as Nix Olympica), rises to a full 15 miles above the general surface level, but others are hardly inferior. Other regions were heavily cratered, and there were also some depressed basins, of which the most impressive were Hellas and Argyre — both visible from Earth, though their nature had not been realized. Hellas had been regarded as an elevated, snow-

covered plateau rather than as a basin.

The Tharsis volcanoes seemed to be associated with drainage systems, and there were many features that could hardly be regarded as anything but dry river beds. Therefore, the Martian climate must have been much less hostile in the past than it is now, though the cause of the climatic variations is still unknown.

The Vikings

The next stage in our exploration of Mars was to attempt a soft landing with an automatic probe. In 1971 the Russians made two such experiments, but in each mission contact with the lander was lost before any useful information could be obtained, and the first successes were delayed until 1976, with the American Vikings.

There were two Viking probes. Each consisted of an orbiting section together with a lander, and the touchdown procedure was purely automatic. Lander 1 came down in the ochre plain of Chryse, while Lander 2 made its descent into another plain, known as Utopia. Both sites proved to be rock-strewn; there was evidence of the past action of running water, and the temperatures were

The surface of Mars
Above On the distinctly red globe as seen from the orbiter of Viking 1, four of the largest craters are clearly visible, the topmost being Olympus Mons.

Right Scene on Mars soon after the first manned landing has been made. The dominant color is orange-red; even the sky is of this hue, because the sunlight is reddened by the quantities of fine dust suspended in the tenuous atmosphere.

very low — below −22° Fahrenheit.

One of the main objects of the Viking program was to search for any trace of Martian life. Samples were collected by means of an extended scoop and drawn back into the main body of the lander, where they were analyzed. To the disappointment of all the investigators, no signs of living matter were found, though it is still too early to dismiss Mars as being completely sterile.

Against this, the Orbiter sections of the Vikings established that the polar caps are made up of water ice, and may be extremely thick. There is in fact plenty of water on Mars, both in the caps and in the surface rocks. This is an encouraging sign for the planners of expeditions there.

Martian Bases

Establishing a base

Because a journey to Mars is bound to take weeks, even after nuclear-powered rockets have been perfected, any manned expedition will have to consider setting up a full-scale base immediately. Quick reconnaissances, such as the Apollo ventures to the Moon, will not be practicable. No doubt the first journeys will be made within the next hundred years — perhaps much sooner.

When the first astronauts reach Mars, they will find much that is unfamiliar. The atmosphere is so thin that fully pressurized suits will have to be used all the time, except inside a spacecraft or the base; the old idea of walking about in the open with no protection other than an oxygen mask has no foundation in fact. On the other hand, the alternation of day and night will not seem strange; a Martian sol is only about half an hour longer than a terrestrial day.

Analysis of the surface rocks has already shown that there is nothing surprising about them. The most abundant elements are silicon and iron, with lesser amounts of aluminum, calcium, titanium and other substances. Moreover, there is every hope that water may be extracted from the rocks.

Much will depend upon whether appreciable quantities of ice exist below the surface, which is by no means unlikely. This raises the question of whether life has ever appeared on Mars during the periods of relative warmth.

The changing climate

One explanation for the comparatively dense atmosphere which formerly must have existed involves the polar caps. The caps are composed of water ice overlaid seasonally by a thin layer of solid carbon dioxide. The axial inclination of Mars varies to a greater extent than that of the Earth; it may become as much as 35 degrees, while at other times it is no more than 14 degrees. There may be periods when both caps melt annually, and in this case the volatiles would be released into the atmosphere, producing a marked increase in density. Conditions would then be favorable for the emergence of life.

The objection is that life is always slow to evolve, and on Mars it might not have had time to develop before the atmosphere became thin once more and all liquid water disappeared. Yet the features shown in the Mariner and Viking pictures clearly resemble well-worn river beds that may have been formed over some considerable time.

One important task for the first expedition will be to examine the Martian rocks to see whether there are any traces of fossils. The astronaut in the foreground is collecting rock samples from the inner wall of a crater near the Martian north pole, from which most of the frozen deposit has evaporated. If fossils are discovered, we will know that Mars has not always been sterile, and in the future, with the changing of the axial inclination, the climate may again become less hostile.

We cannot yet decide upon the form of a permanent base, but the science-fiction concept of a hemispherical dome, kept inflated by the air pressure inside it, cannot be ruled out. At least there will be no danger from alien beings; the atmosphere, useless though it is for breathing, is not toxic; and there will be no strong ground tremors.

Transport on Mars

If several bases are set up at about the same time, communication will be all-important. Mars must be almost a silent planet. Sound would be very feebly propagated in the tenuous atmosphere. Beyond the planet there will be no difficulty in keeping in touch with Earth, except when the two planets are on opposite sides of the Sun (the Vikings have shown that), but radio links between two widely separated stations on Mars itself may be more troublesome, because the ionosphere is not sufficient to reflect radio waves back to the ground and make direct long-range signalling possible. Phobos and Deimos may be useful as relays, but neither satellite is visible from very high latitudes, and it may be more practicable to use artificial satellites.

For transport, wheeled vehicles can no doubt be used; the surface of Mars is no more rocky and uneven than that of the Moon, at least in the ochre plains, but no aircraft of conventional type will be of any help at all. On Earth, passenger aircraft cannot operate at heights of over 100,000 feet above sea level, and even at this altitude the air pressure is greater than it is on the surface of Mars.

The astronaut in the background of the painting has just taken off on a one-man flying device, which would give greater mobility than ground transport. The 'platform' could probably be used up to a range of about 30 miles.

The chances of persuading any Earth-type plants to grow in the open are slim. The temperatures are too low at night, and moreover the atmosphere is a poor shield against harmful radiations coming from space. Yet the Martian colony must be as self-supporting as possible; to bring essential materials from Earth will be much more difficult than to take them to the Moon, because of the greater distance involved.

Mars in the future

If it is decided to set up permanent colonies on Mars, there are many problems to be faced. Every natural resource must be exploited to the full, and the colonists must be carefully chosen; any major conflicts would clearly be disastrous. There must be men and women of all skills and from all walks of life; in particular, a medical service will have to be set up immediately.

In the long term, there may be additional complications. The surface gravity on Mars is low: only about one-third that of the Earth. There is no reason to suppose that this will be harmful, but we must remember that even at an early stage in the existence of the colony, children will be born, and we cannot be sure that a boy or girl who has spent years on Mars will be able to adapt to the stronger pull of the home planet. We may even reach a stage in which there are 'Martians' who can never hope to visit Earth at all.

If this is so, then it is a situation which must be accepted; but it may cause a well-defined division in the human race, perhaps even with obvious physical differences. Looking still further ahead, can we entirely discount the possibility of a planetary conflict — not between Earthmen and aliens, but between Earthmen and the 'new Martians?' Human nature is slow to change, and there is, unfortunately, no reason to suppose that transfer to Mars will produce a rapid improvement.

The ideal would be to alter the planet itself, and make it more nearly like the Earth. Nothing can be done about the low gravity, but science-fiction writers have made great play of the possibility of providing Mars with a breathable atmosphere. Our present technology is quite unequal to the task, but it would be dangerous to claim that this will always be so. One major objection is that even if Mars were given such an atmosphere, it could not retain it for long on the cosmical time-scale, so that it would have to be periodically renewed. But all this is highly speculative, and for the moment at least we must accept Mars as an unfriendly world.

Moons of Mars

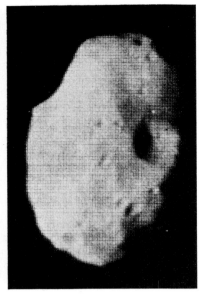

Phobos *above*
First close-up views of the inner Martian moon Phobos: Left from 9175 miles (14680 km.); right from 3460 miles (5540 km.)

One of the most interesting features of Mars is its pair of satellites. Phobos and Deimos (named after the two mythological attendants of Mars, the War God) were discovered in 1877 by the American astronomer Asaph Hall. They had not been seen before because both are very small and faint, so that powerful telescopes have to be used in order to show them.

Both satellites revolve virtually in the plane of the planet's equator, which is why they are never visible from the polar regions of Mars. Also, both are close to the ground. Phobos moves at a mean distance of only 5,800 miles from the center of Mars, which means that it is only 3,700 miles above the surface; the distances for Deimos are 14,600 miles and 12,500 miles respectively. The revolution periods are 7 hours 39 minutes for Phobos and 30 hours 18 minutes for Deimos.

The Martian 'month'

Since Mars has a rotation period of 24 hours 37 minutes, the Martian 'month', as reckoned by Phobos, is shorter than the day! To an observer on the planet, Phobos would seem to rise in the west, cross the sky in a direction opposite to that of the Sun and stars, and set in the east only 4¼ hours later, during which time it would go through more than half its cycle of phases from new to full. The interval between successive risings would be only a little over eleven hours. Moreover, Phobos would seem appreciably larger when high up than when low down. Often it would be eclipsed by the shadow of Mars, and only near midsummer and midwinter would it be free of shadow for the whole of its passage across the sky.

Deimos behaves differently. As Mars spins, Deimos almost keeps pace with it, and as seen from one site on the planet's surface Deimos would remain above the horizon for over 60 consecutive hours, passing twice through its full cycle of phases. Its eclipses would be similar to those of Phobos, though not quite so frequent.

Phobos is unique in the Solar System in that its revolution period is less than the rotation period of its primary. Some years ago, calculations were made indicating that it was spiraling slowly downward, and would hit the planet in 50,000 years from now. From these calculations, the Russian astronomer Shklovskii suggested that Phobos must be being 'braked' by friction with the extremely thin upper part of the Martian atmosphere. This would mean that the mass of Phobos must be negligible, as otherwise the frictional braking would be too slight to be noticed. He went on to propose that Phobos might be nothing more nor less than a hollow space-station launched by the Martians!

Not surprisingly, this theory met with little support, and before long it became clear that the original calculations were wrong; Phobos has a stable orbit and is not spiraling downward. All the same, it is a strange little body, and it began to seem even more curious in 1969, when a close-range photograph from Mariner 7 showed that it is irregular in shape.

Mariner 9 and the Vikings

A much closer-range picture was obtained from Mariner 9 in November 1971, which was subsequently followed by the Viking orbiters in 1976, giving us the best pictures to date. The shape is very irregular, though basically elliptical; the average dimensions are, as expected, about 16 miles by 13 miles. The largest crater shown is over 4 miles across. Whether it is a blowhole crater of internal origin, or whether it is due to a meteoritic impact, is not yet certain.

Phobos looks very much like a 'bit of something' – and this may well be the case. Quite possibly it is not a true satellite, but merely a member of the asteroid belt which came close to Mars in the remote past and was captured. This would admittedly involve some coincidences, particularly since there can be little doubt that Deimos is of the same nature; but at least we may be sure that both dwarf attendants are basically different from our own massive Moon, or from the larger satellites of the giant planets. Evidently it is rocky and very old, and it must have considerable structural strength.

Natural space stations

Suggestions have often been made that Phobos (and Deimos, too) could be pressed into service. The satellites might be regarded as natural space-stations. Their gravitational pulls are very slight; the escape velocity of Phobos is about 30 miles per hour, and that of Deimos rather less. In theory, this is still too high to allow an astronaut to 'jump clear' of the satellite by muscle power alone; but anyone who leaped upward from Phobos or Deimos would take a very long time to come down, and could well be classed as a temporary independent satellite of Mars!

It is not likely that men will land on either Phobos or Deimos before setting foot on Mars itself. There would be nothing to be gained by making the attempt, which would be extremely difficult from a navigational viewpoint. However, we may be assured that touchdowns will be made there eventually. It will not be a case of a conventional landing maneuver; it will be much more in the nature of a docking operation, since the tiny gravitational pull of the satellite will be to all intents and purposes inappreciable.

From Deimos, the view of Mars will be magnificent; it is shown in the painting opposite. Mars is seen at half-phase, so that its eroded craters are best seen along the terminator (that is to say, the boundary between the sunlit and the night hemispheres). At the top is the bright south polar cap. Toward the lower right is one of the most interesting features on Mars, the object long known as the Olympus Mons. It is visible from Earth with a moderate-sized telescope when Mars is well placed; it then appears as a small round spot. When photographed from Mariner 9, in 1971, it was found to be a giant 15-mile volcano larger than any of the famous craters of the Moon. Other craters photographed by the Mariners and Vikings are shown.

As seen from Deimos, Mars would appear to spin very slowly – because, as we have seen, the rotation period of Mars is not much shorter than the period taken by Deimos to complete one circuit of the planet. Closer in, from Phobos, Mars would dominate much of the sky, and would cast a brilliant reddish light on to the irregular rocks and pits of the satellite.

Though we are looking far into the future, the possible use of Phobos and Deimos as space-platforms is worthy of closer consideration. Judging from the Viking photographs, they are firm, solid bodies with considerable cohesion, and even though their escape velocities are so low there are some scientific experiments in which a very slight pull of gravity might be useful. Also, there is the question of communication. Almost certainly it will be necessary to establish relay satellites round Mars soon after the first bases are set up there (except in the unlikely event of the planet proving to have a useful ionosphere), but a radio station on Deimos, in particular, would be extremely valuable inasmuch as it could give coverage over a full hemisphere of the planet.

The meteorology of Mars

There is, too, the question of the meteorology of Mars. Wind systems occur; moving clouds can be observed even from Earth observatories, and it seems that the dust storms which occasionally hide the surface must be wind-raised. From Deimos, weather systems could be observed in their entirety, and closer-range studies could be carried out from Phobos, which may not lie very far above the detectable limits of the Martian exosphere (that is to say, the outermost layer of the planet's tenuous atmosphere).

It remains to be seen whether scientific bases on the natural satellites will have any advantages over orbital bases of the Skylab type; but the possibilities are there, and may one day be exploited.

It will also be of the greatest interest to analyze the material of Phobos and Deimos themselves. At present their densities are unknown (the escape velocities given above have been calculated on the assumption that the densities of the satellites are equal to that of Mars itself, which appears to be reasonable). It is not likely that Phobos and Deimos ever formed part of Mars. Either they were produced by material which was once spread round the planet, or else they are captured asteroids. Probably it will be possible to reach them before we can send our probes out into the belt of asteroids, and so they may provide us with our first samples of material from the outer part of the Solar System.

Whether or not we establish relay stations or scientific bases on Phobos or Deimos, it is difficult to doubt that Mars itself will be reached in the foreseeable future. The paintings shown here represent what we believe will be found there; and thanks to the Viking probes we have a great deal of reliable information to guide us. But we must always remember that Mars has given us plenty of surprises in the past, and may have many more stored up for us in the years to come.

Perhaps we may look back to the words of Percival Lowell, written in 1906. He may have been wrong in his interpretation of the so-called Martian canals, but at least he put forward an idealistic view of the attitude of his 'Martians', whom, he believed, had outlawed warfare and had united in order to make the best of their arid world. There could be no conflict upon Mars. In Lowell's words: 'War is a survival among us from savage times, and affects now chiefly the boyish and unthinking element of the nation. The wisest realize that there are better ways of practising heroism and other and more certain ends of ensuring the survival of the fittest. It is something people outgrow.' Let us hope that we, too, have outgrown it before we set up the first base upon the red deserts of Mars.

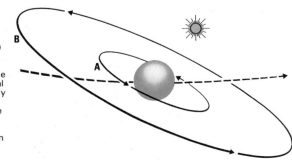

Phobos and Deimos
The orbits of Phobos (A) and Deimos (B). Both satellites move almost exactly in the plane of the planet's equator; the axial inclination of Mars is only slightly greater than that of the Earth. Because the satellites move in the equatorial plane, an observer near the Martian poles would never be able to see them.

Mars as seen from its outer satellite *right*
Mars is at half-phase, so that the large craters along the terminator are partly shadow-filled and are well seen. The south polar cap, now known to be a thick layer of water ice, appears to the top of the picture; Olympus Mons is to the lower right. Near the edge of the 'night' half of Mars is Phobos, the inner satellite, which is much closer to Mars. The two dwarf satellites can never approach each other to within a distance of much less than 9,000 miles.

Venus: Hell Planet

Venus, the brilliant planet which shines so gloriously in our skies, has often been nicknamed 'the Earth's twin'. In size and mass it is strikingly similar to our own world.

Before the Space Age it was thought that Venus might be a reasonably friendly world, with plenty of water on its surface and a tolerable temperature. Unfortunately this is not so. Probes have shown that the surface is fiercely hot, that the atmospheric pressure is crushing, and that the clouds contain quantities of sulphuric acid. Life there appears to be absolutely out of the question, and neither are there any real prospects of manned expeditions. It is strange to reflect that less than two decades ago Venus was regarded as a more promising potential colony than Mars.

The veiled planet *left*
During its flyby of Venus in February 1974 Mariner 10 sent back the first close-range pictures of the upper clouds. In this view the banded structure is clearly shown, and the atmospheric circulation near the poles is quite different from that near the equator. Successive photographs confirmed the relatively quick rotation of the cloud tops. Only one flyby was made, since the main target of Mariner 10 was the innermost planet, Mercury.

Through an Earth-based telescope Venus shows a brilliant, almost blank disk. Few features can be seen, and those that are visible are always vague and ill-defined. The surface of the planet is permanently hidden by the dense, cloudy atmosphere, and before 1962 very little was known about the surface conditions. Venus was aptly termed 'the planet of mystery'.

There were two main theories. Many astronomers regarded Venus as a mainly ocean-covered world, with clouds made up of water; alternatively it was suggested that the surface was too hot to permit the existence of liquid water, which would leave Venus a hostile dust-desert. Even the length of the rotation period or 'day' was not known, though it was tacitly assumed to be equal to about four terrestrial weeks. Spectroscopic examination had shown that the atmosphere must be made up largely of carbon dioxide, but water vapor was suspected also.

Inevitably there was speculation about the possibility of life. It was pointed out that Earth life began in the seas at a time when the temperature of the world was relatively high, and when the atmosphere was rich in carbon dioxide; if, then, Venus were ocean-covered, it was by no means unreasonable to expect that life there would have started to evolve in comparable fashion. The planet could, in fact, have been in a 'Carboniferous' condition, in which case the existence of amphibians or even primitive reptiles could not be ruled out.

The first probes

This attractive picture was shattered in 1962, when the first successful planetary probe, Mariner 2, bypassed Venus and sent back reliable information. The surface temperature proved to be extremely high, and the 'marine theory' had to be abandoned. The rotation period was found to be about 243 days; and since Venus takes almost 225 days to go once round the Sun, the 'day' is longer than the 'year.' To make things even more peculiar, Venus has a retrograde or wrong-way rotation, so that in theory an observer on the surface would see the Sun rise in a westward direction and set toward the east; the length of a 'solar day' there is 118 Earth days. The reason for this curious behavior is unknown.

Subsequent radar work, carried out in America, provided the first rough map of the surface, and it was established that there are large, rather shallow craters. Meantime the Russians had started to dispatch probes of their own, and in December 1970 Venera 7 made a controlled landing, coming down through the dense atmosphere by means of parachute and transmitting for almost an hour after arrival before being put out of action by the hostile conditions. In 1975 the Soviet space planners achieved perhaps their most surprising triumph yet when they soft-landed two more probes, Veneras 9 and 10, and obtained one picture from each. The light-level on the surface of the planet was surprisingly high (the Russians compared it with that in Moscow at noon on a cloudy winter day), and there were rocks everywhere. Windspeeds were gentle — well below 10 mph — but in an atmosphere so dense as that of Venus, even a moderate breeze will have great force.

The deadly rain

The first close-range pictures of the cloud-tops were obtained by the American probe Mariner 10 in February 1974. Only one pass of Venus was made, since the Mariner was on its way to Mercury, but the pictures were spectacular, and they confirmed that the rotation period of the upper clouds is a mere four days; the atmospheric structure must be extremely strange.

The clouds are totally unlike the clouds of Earth, and all the evidence suggests that they contain appreciable quantities of sulphuric acid. They may well produce 'rain,' but this rain will consist of sulphuric acid droplets, and even though it may not penetrate as far as the actual surface it poses an obvious threat to future probes. We also know that the surface temperature is of the order of 900 degrees Fahrenheit, and that the ground pressure is at least 90 times that of the Earth's air at sea level. The more we learn about Venus, the more unwelcoming it seems to grow.

Why is Venus so different from the Earth? The reason must lie in the lesser distance from the Sun. The surface was never able to cool down, and eventually there was what may be termed a 'runaway greenhouse' effect; carbonates were driven out of the surface rocks, producing the thick carbon dioxide atmosphere of today.

Carl Sagan, an American astronomer, has suggested that in the future it may be possible to 'seed' the atmosphere of Venus and split up the carbon dioxide molecules, releasing free oxygen and providing the planet with an air more similar to our own. This may become practicable in the far future, but our present technology is unequal to it. For the time being Venus is a world to be examined from a respectful distance.

Into the future

One thing we do not yet know is whether there is active vulcanism going on now. There may well be; radar maps of the planet are still very rudimentary, and all we have to guide us is the limited information drawn from Veneras 9 and 10. We must await the results of future research. The Americans have not made any definite plans for automatic soft landings, though a Pioneer probe to be launched in 1978 should send back further information during its descent through the atmosphere.

In some ways Venus has shown itself to be a disappointment. We cannot conceive of any life that could survive there, and to attempt a manned landing would be to court disaster. The 'Planet of Love' is less inviting than it looks; we must regard it as the most hostile of all the worlds in the Sun's family.

The phases of Venus *right*
Since Venus is closer to the Sun than we are, it shows lunar-type phases. These photographs, taken by H. R. Hatfield, show Venus at half-phase (right) and crescent (far right). When Venus is closest to us, its dark side is turned toward us; when full, it is virtually behind the Sun.

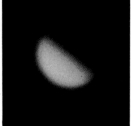

A glimpse of the surface *below*
Only two pictures have so far been obtained from the surface of Venus. The first of these, obtained from Venera 9 on October 12, 1975, is shown here. The descent through the atmosphere was automatic, and before entry the probe had been chilled so as to protect it from the intense heat as efficiently as possible. Rocks are everywhere; part of the Venera is shown at the bottom of the picture. Venera 9 transmitted for approximately one hour. Venera 10, which landed a few days later, confirmed the main findings, and showed a landscape of the same general type.

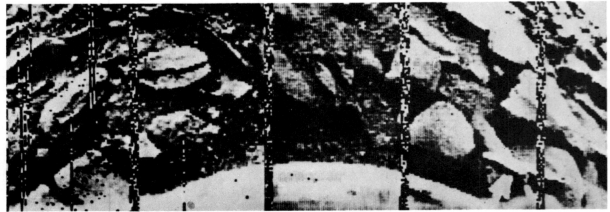

Venus and Earth compared *left*
On this scale the difference between the two planets is inappreciable. In fact, Venus is very slightly smaller (diameter 7,700 miles as against 7,926). The mass of Venus is 0.81 of that of Earth, its volume 0.92 that of the Earth and its density 4.99 (where water equals 1). If Venus and the Earth are so alike in size, mass and density, why are they so different in character? Venus is closer to the Sun and therefore its higher temperature has caused the planet's whole development to be different. Venus' distance of 67,200,000 miles from the Sun has ensured that the planet, often nicknamed 'the Earth's twin,' can never produce life.

A deadly world *right*
A hydrogen-filled balloon floats above the gloomy surface of Venus, sending back information about this strange, possibly dynamic world. There is a high volcano, still active and with its upper crater glowing redly in the half-light; the tangled rocks are strewn everywhere, suffering perpetual erosion from the slow but immensely powerful winds. Above, the dense clouds hide the Sun; from Venus the sky itself can never be seen, since the clouds never draw back to give a view of the universe beyond. The scene conjures up the impression of a Dantéan inferno, and it is easy to see why Venus has been nicknamed 'the hell planet.'

Mercury: Elusive World

Mercury, the innermost planet, is a small world. Its diameter is only about 3,000 miles (slightly less, according to the latest measurements), so that it is not a great deal larger than the Moon. Its atmosphere is negligible, as is only to be expected in view of its low escape velocity and its nearness to the Sun. It is much less easy to observe than Venus, since it always remains comparatively close to the Sun in the sky.

Maps of the surface made before 1974 were very rough, but in that year the American probe Mariner 10 made the first of its three active passes of Mercury, and sent back detailed pictures. Mercury proved to have a surface superficially very like that of the Moon, with lunar-type features. A very tenuous atmosphere was detected, and it was also established that Mercury has a weak but appreciable magnetic field. Less than half of the total surface was studied from Mariner 10, but there is no reason to assume that the rest is essentially different.

Mercury and Earth compared *above*
Mercury (diameter 3,000 miles) is much the smaller and less massive of the two; its low escape velocity means that it has been unable to retain an atmosphere.

Before the Space Age, Mercury was a somewhat neglected planet, chiefly because it is so difficult to study from Earth. The best pre-Mariner map of the surface was drawn up by E. M. Antoniadi in the 1930s. Antoniadi used the great 33-inch refractor at the Observatory of Meudon, near Paris, but even so he could do no more than show some darkish patches and brighter areas; and we now know that his chart was not even approximately accurate. It was only with the flight of Mariner 10, in 1974, that

lies at a 'hot pole' (hence its name). There are few lunar-type maria, but the craters are everywhere, and seem to obey the same laws of distribution as those of the Moon, so that they were presumably formed by the same process. Ray craters are also seen. One of these, the first feature to be positively identified from Mariner 10, was named 'Kuiper' in honor of Gerard P. Kuiper, who played so important a role during the early days of probe exploration and who had died some time before the Mariner 10 mission.

The Mercurian scene
Mercury is an unfriendly world, but in some ways it is less hostile than Venus, and there is no reason why future expeditions should not land there. Yet the difficulties will be immense. We cannot expect to find any useful materials; all the evidence suggests that the crust of Mercury is very similar to that of the Moon.

The scene shown in the painting is probably very near the truth. A solitary plug of lava, ejected from a volcanic vent many millions of years ago, forms the only true landmark apart from the scattered craters. Everything spells

'desolation'; the loneliness and the silence are absolute. In the sky, the Sun – here hidden by the rock – is terrifying; on average its disk seems three times larger than as seen from Earth, and by shielding his eyes from the surrounding glare the observer will be able to see both the solar corona and the glow of the Zodiacal Light, which is caused by thinly spread material in the main plane of the Solar System.

We cannot tell when astronauts will reach the planet, but there is every reason to suppose that automatic probes will be landed there in the foreseeable future. Their main function would be to study the Sun from relatively close range.

The rotation of Mercury
In one important respect our ideas about Mercury have changed since the start of the Space Age. It used to be thought that Mercury's 'day' and 'year' were the same length: 88 Earth-days. Had this been the case, Mercury would have kept the same face turned permanently toward the Sun, and would behave in the same way as the Moon does with respect to the Earth. There would have been a zone of permanent

day, an area of everlasting night, and an intermediate 'twilight zone', over which the Sun would have moved alternately just above and just below the horizon. It was only in the 1960s that radar measurements proved this picture to be wrong. In a way, the discovery made things more difficult. The temperatures in the so-called twilight zone would have been tolerable, with no fierce extremes of either heat or cold. Unfortunately, we now know that there is no twilight zone; every part of Mercury experiences both day and night, though the calendar is strange by our standards.

From Mercury, the view of the Solar System will be rather restricted. Like Venus, the planet has no satellite; and neither is there a planet still closer to the Sun, though a century ago astronomers believed in the existence of such a planet and even gave it a name – Vulcan. Occasional comets will come into view, their tails stretching across the blackness of the night sky; Venus will be brilliant, and the Earth-Moon system will appear as a bright double star. Yet nothing can alleviate the loneliness and the bleakness of Mercury – a world that has never known the breath of life.

we learned anything positive about the surface features.

Because Mercury is so close to the Sun, it experiences great extremes of temperature. It has a rotation period of $58\frac{1}{2}$ days, and a 'year' of 88 Earth days; to an observer on the planet the interval between one sunrise and the next would be 176 Earth days, though the variable orbital speed and the slow rotation would make the Sun move somewhat erratically in the Mercurian sky. The maximum temperature at a 'hot pole' is of the order of 700 degrees, but the nighttime temperature is low, because there is no atmosphere dense enough to waft appreciable warmth from the day to the night hemisphere.

Though Mariner 10 flew past Mercury three times before its power failed, it photographed almost the same regions on each occasion. In particular it showed only half of the Caloris Basin, which

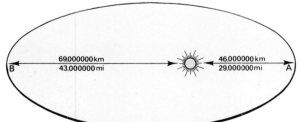

B	69,000,000 km		46,000,000 km	A
	43,000,000 mi		29,000,000 mi	

The orbit of Mercury
Above The orbit of Mercury, which is relatively eccentric; the distance from the Sun ranges between 29,000,000 miles and 43,000,000 miles. The revolution period is 88 Earth-days.

Left Because of its eccentric orbit, the Sun as seen from Mercury appears considerably larger when the planet is near perihelion. The diagram shows the Sun as seen from Mercury, to scale, at perihelion and at aphelion.

Approaching the planet *above*
Three photographs of Mercury, taken from Mariner 10 in March 1974 when the probe was 591,900 miles from the planet. The diameter of the smallest object shown on the last picture is 12.4 miles. Craters show up best when close to the terminator.

The surface of Mercury *right*
Mercury is a lifeless world of silence and desolation. The only landmark in this scene is a plug of lava, sent out by a volcanic eruption millions of years ago and since worn down by the alternate expansion and contraction of the surface materials owing to the great range of temperature between day and night; but no volcanic eruptions happen on Mercury now. Mercifully, the glaring disk of the Sun, three times the size that it appears from Earth, is hidden by the rock column. The Earth-Moon system can be seen in the black sky, immersed in the band of the Zodiacal Light produced by debris spread in the main plane of the Solar System. From the hidden Sun, the pearly corona stretches out toward the zenith.

Asteroids: The Minor Planets

Between the orbits of Mars and Jupiter, separating the two main groups of planets in the Solar System, we meet the swarm of dwarf worlds known as minor planets, planetoids or asteroids. Several thousands of them are known, and their total number has been estimated at anything between 44,000 and 100,000; but most of them are tiny, irregular bodies. Only three known asteroids have diameters of over 250 miles; these are Ceres (650 miles), Pallas (355) and Vesta (335). No asteroid is massive enough to retain any vestige of atmosphere.

It has been suggested that the asteroids represent the remnants of a former planet (or planets) which met with some disaster in the remote past, and disintegrated. Alternatively, it may be that the swarm was formed from material that never condensed into a planet; but all the asteroids put together would not make one body nearly so massive as our Moon.

Asteroids are often regarded as rather erratic members of the Sun's family. Though they all move round the Sun in the same sense as do the Earth and the main planets, some of them have orbits which are highly eccentric and inclined; Pallas, the second largest of the entire swarm, has an orbital inclination of over 34 degrees.

Though we have no information about their surface features, it is logical to suppose that the larger asteroids are at least approximately spherical. This is not true of some of the smaller members of the swarm, particularly those with exceptional orbits which swing them away from the main group and bring them relatively close to the Earth.

Eros

The first and best-known of this group is Eros, which was discovered in 1898. It has a period of 1¾ years, and its orbit swings it from just outside that of the Earth to far beyond Mars. At its nearest to us, it can come to a distance of only 15,000,000 miles – as it did in 1931, and more recently in 1975. Variations in its brightness show that it is irregular in shape, and it has even been seen elongated. Apparently Eros is shaped like a cigar 18 miles long and only about 4 wide.

Since it wanders about from the inner part of the Solar System well into the asteroid zone, Eros might be regarded as an excellent site for a 'space beacon' – and the painting shows a future scene as an expedition lands on the curious little world. Of course the small size and irregular shape produce remarkable effects of tilt; the force of gravity is negligible and, as with the satellites of Mars, touch-down will be more in the nature of a docking operation than a landing maneuver.

The sky as seen from an asteroid would be unfamiliar. Eros, of course, would provide great variety; when near perihelion the view would be much as seen from Earth (allowing for the lack of atmosphere), but when Eros is at its furthest from the Sun there will be many asteroids visible with the naked eye to any human observer who happens to be there. The rotation period is a mere five hours, so that the sky will spin round relatively quickly.

From a world in the thick of the main swarm, the view will be dominated by asteroids. Some of them will show visible disks, crater-pitted and irregular; others will appear as faint stars; there will be obvious movement, and the sky will be anything but static. Now and then one of the larger members will pass by, looming large for a few hours or days before receding once more; and there is always the possibility of collision.

Icarus

Of all the asteroids, perhaps the most extraordinary is Icarus. Icarus is unique in that it passes within the orbit of Mercury. At its closest to the Sun it is only 17,000,000 miles from the solar surface; at its aphelion only about 200 days later, it has moved out to a distance of 183,000,000 miles – well beyond Mars. Icarus must have the most uncomfortable climate in the Solar System.

When at its closest to the Sun, Icarus must glow a dull red as its rocks are intensely heated. Since it spins round once in every 80 minutes, the surface can hardly have time to cool before it is again bathed in the full glare of the solar radiation. In the painting Icarus is shown near perihelion; in the background the Sun blazes furiously, the pearly corona streaming out until it envelopes the wandering asteroid. The surface of Icarus is cracked and pitted, and yet only 200 days later it will be bitterly cold, with the Sun shrunken and pale in the distance.

When Icarus passes by the Earth, it

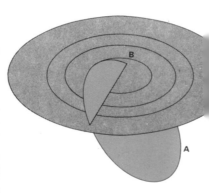

The orbit of Icarus *above*
The revolution period of Icarus is 409 days, and the distance from the Sun ranges between 183,000,000 miles at aphelion to only 17,000,000 miles at perihelion. The orbital inclination is 23 degrees. 'A' is the orbit of Icarus; 'B' that of Mercury.

might be possible to dock with it and set up an automatic transmitting station on its surface. The value of such a station could not be denied because it would provide a superb opportunity for studying the Sun from close range; but it is problematical whether any scientific instruments could survive under the conditions existing on Icarus at perihelion. The only hope would be to put them under the surface, which would reduce their value immediately. But the scene as it flew past the Sun at only 17,000,000 miles defies description; its glowing surface, the solar blaze and the contrasting blackness of the sky make up a picture which is beyond anything in our wildest dreams.

Icarus *above*
Sweeping in from beyond the orbit of Mars, the tiny asteroid Icarus passes only 17,000,000 miles from the Sun, closer even than Mercury. As it does so, its surface is so intensely heated that it must glow red-hot.

Arthur Clarke once suggested that in its cool cone of shadow, or well dug-in beneath its surface, scientists might use Icarus to put themselves and their equipment close to the Sun, shielded by its 10,000-million ton bulk.

The position of Vesta
Above left Vesta, brightest and third largest of the asteroids, photographed by F. J. Acfield in 1967. Vesta appears as a starlike point between the arrows. The cross above shows the position of Vesta twenty-four hours later. As will be seen, Vesta's appearance is exactly like that of a star.

Left Two sketches of the position of Vesta made by Patrick Moore in 1969. The interval is 24 hours. The stars remain in the same relative position, but Vesta has moved.

The Eros Expedition *right*
Astronauts have docked with the asteroid, and are setting up an inflatable, semi-transparent dome; they are preparing to make a geological survey of Eros itself. The asteroid is 18 miles long by 4 wide, and spins round in 5 hours. Its surface is pitted with craters, due mainly to collisions with cosmical debris in the asteroid belt; near its aphelion point Eros enters the main zone, though when at its closest to the Sun it approaches the orbit of the Earth, and may come within 15,000,000 miles of us.

Jupiter: Killer Planet

Jupiter, innermost of the giant planets, is a world as different from Earth or Mars as it could possibly be. Its huge globe, over 88,000 miles in diameter as measured through the equator, is clearly flattened; this is due to the rapid rotation, since the Jovian 'day' is less than 10 hours long. The outer layers are made up of gas, mainly hydrogen and unprepossessing hydrogen compounds such as ammonia and methane. The internal constitution of Jupiter is not certainly known; it was once thought that there might be a rocky core, surrounded by a thick layer of ice which was in turn overlaid by the deep, hydrogen-rich atmosphere; but it is now believed that most of Jupiter is made up of liquid hydrogen, with a gaseous surface and a relatively small rocky core. The internal temperature must reach 54,000 degrees Fahrenheit (30,000 degrees Centigrade), but there is no suggestion that Jupiter is stellar; the core heat is much too low for nuclear reactions to take place.

Man can never land on Jupiter. The surface is not solid; the gravitational pull is extremely high, and the planet is surrounded by zones of radiation that would be lethal to any astronauts venturing inside them. The most we can hope to do is to come down upon some of the members of the satellite family – such as the innermost, Amalthea.

Although Jupiter has thirteen satellites; a fourteenth is suspected. The outer members of the family are small and distant, so that they are probably captured asteroids, but the four largest satellites – Io, Europa, Ganymede and Callisto – were discovered soon after the invention of the telescope. Europa is slightly smaller than our Moon, Io slightly larger, and Ganymede and Callisto much larger; indeed, the diameter of Ganymede is greater than that of the planet Mercury.

Pioneers to Jupiter
The first Jupiter probe was Pioneer 10, which bypassed the planet in December 1973. It obtained pictures from comparatively close range, and sent back other information, some of it distinctly surprising. The Great Red Spot, a vividly colored oval area 30,000 miles long by 8,000 miles broad, proved to be a kind of whirling storm – a phenomenon of Jovian meteorology, rather than the top cratering). More may be expected from the Voyager probes, which were launched in 1977 and are scheduled to fly past Jupiter in 1979, before going on to rendezvous with Saturn.

Io
Io, the innermost of the large satellites, is a remarkable world. It has a definite effect upon the Jovian radio emission, so that clearly it moves well inside the

Jupiter from Europa *above*
Jupiter as seen from Europa, second of its four large satellites. We are standing near the north pole of Europa; the fore-ground is in the shadow of the hills behind us, but sunlight is striking the mountains, which are covered with 'ice' – not ordinary ice, but gases which have been frozen; the temperature is below −200 degrees Fahrenheit.

Radio emissions *below*
A speculative diagram of some of the characteristics of the radio emissions from Jupiter.
A Circular-polarized radiation
B Plane-polarized radiation
C Jupiter's rotation axis
D Jupiter's magnetic axis
E Magnetic field lines
F Paths of trapped vehicles

of a column of stagnant gas or a floating island, as had previously been believed. The bright zones on the disk were relatively high and cold, whereas the dark cloud belts were areas in which the gas was descending.

As expected, Jupiter has a very powerful magnetic field, which is associated with the radio waves emitted by the planet. There are also surrounding zones of intense radiation, which very nearly put the Pioneer 10 instruments out of action. Pioneer 11, which followed a year later, was put into a rather different path, so that it passed relatively quickly over the equatorial regions of Jupiter, where the radiation is strongest.

Europa
The painting on this page shows the scene from Europa, which orbits at a distance of 417,000 miles from the center of Jupiter. The satellite has a reflective surface, and is extremely cold; there is virtually no atmosphere, so that the sky is black.

Both the Pioneers attempted to obtain pictures of the major satellites, but without much success (though Ganymede at least shows indications of

planet's magnetosphere. It is thought to have a salty surface; the crust is being constantly bombarded by the radiation, producing a tenuous cloud of sodium – an 'atmosphere' unique in the Solar System. No expeditions to Io are likely to be dispatched, either now or in the future, because of the radiation danger. Io must in fact be a deadly world.

Amalthea
Even closer to Jupiter there moves Amalthea, a tiny world no more than 200 miles in diameter at most. It was discovered in 1892 by the American astronomer E. E. Barnard, and is too faint to be seen with small telescopes. It would be a magnificent site from which to study Jupiter, but the radiation means that any exploration will have to be by automatic probes.

Deadly radiations *right*
A Voyager-type probe passes the radiation-soaked surface of Amalthea. Jupiter, only 70,000 miles away, fills more than a quarter of the sky, and the ever-changing atmospheric maëlstrom forms an incredibly beautiful panorama, which, however, no man can observe directly; an astronaut venturing onto Amalthea would be quickly killed by the Jovian radiations.

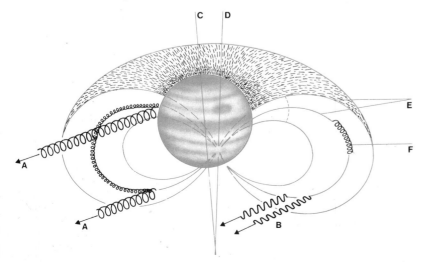

Saturn: Ringed Giant

Beyond Jupiter, moving at a mean distance of 886,000,000 miles from the Sun, lies Saturn – by far the loveliest object in the entire Solar System. Like Jupiter it has a gaseous surface, and is intensely cold; like Jupiter, too, it has a satellite family, in this case made up of ten members. Saturn is smaller than Jupiter, having an equatorial diameter measured at approximately 75,000 miles.

The glory of Saturn lies in its rings, unique in our experience. The rings are composed of small particles, probably ices or at least ice-coated; they may be the remnants of a former satellite that was broken up, or they may be debris which never condensed into a satellite. The rings, as well as the inner satellites, lie in the plane of Saturn's equator.

There can be no hope of landing on Saturn, either with manned or with unpiloted vehicles. As with Jupiter, there is no solid surface. Yet we have a wealth of satellites from which to choose; and the painting shows the view which would be obtained from Rhea, the sixth satellite in order of distance from the planet.

Rhea moves at 328,000 miles from the center of Saturn, so that in this scene the surface of the planet is less than 300,000 miles away. The beautiful ring-system is edge-on, and is visible only as a thin white line fading into darkness to

the right, where the shadow of the planet falls onto it. Yet the rings make their presence known in other ways. The myriads of tiny, icy particles reflect sunlight onto the upper dark half of the planet, and a shadow is cast onto the sunlit portion.

The inner satellites

Four other satellites are visible, three of them giving the impression of pearls on a necklace since they, too, are in the same plane as the rings. The large satellite against the night side of Saturn is Dione, only 93,000 miles closer in; the

others are Tethys, Enceladus and Mimas.

Rhea itself has a diameter which may be as much as 1,000 miles. It is twice as dense as water, and the landscape shown here is basically somewhat similar to that of our Moon. In these remote depths of the Solar System there seems no reason why jagged rocks should not remain unchanged over vast periods of time; the temperature range is not great, since the cold is always so intense, and there may be fewer meteorites to pit the surface with craters. In this scene, the Sun is below the horizon, but the strong yellow light of Saturn casts a brilliant

glow over the bleak, inhospitable rocks of Rhea.

Should it become possible to set up observation posts within the system of Saturn, Rhea will be a strong candidate, since it is reasonably close to the planet and yet is outside the ring zone, where there will be a great deal of potentially dangerous debris. Rhea takes $4\frac{1}{2}$ Earth-days to complete one journey round Saturn, whereas Saturn itself spins round in only $10\frac{1}{4}$ hours, so that the panorama will be quickly-changing, and all areas of the ringed planet will come into view in relatively rapid succession.

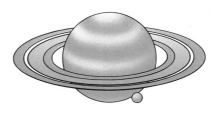

Two aspects of Saturn
Above Saturn as seen from Rhea, the sixth satellite in order of distance from the planet. The rings appear as a thin line; the inner satellites Dione, Tethys, Enceladus and Mimas appear in the black sky. The surface of Rhea is assumed to be not unlike that of our Moon.

Left Saturn and Earth compared. Saturn has an equatorial diameter of 75,100 miles, but it is obviously flattened, and its polar diameter is less than 70,000 miles. It is, of course, a giant world, inferior only to Jupiter; but its density is very low, and is actually less than that of water. Its mass is 95 times that of the Earth, and the escape velocity is 22 miles per second.

Titan: Largest Moon

Of the ten satellites of Saturn, much the largest and most important is Titan, which moves round the planet at a distance of 760,000 miles (reckoned from the center of Saturn; that is to say, rather more than 720,000 miles from the top of the cloud-layer). It has a period of 15 days 22½ hours; its orbit is practically circular, and lies only half a degree in inclination from the plane of the rings.

Titan was discovered by the Dutch astronomer Christiaan Huygens as long ago as 1655, and is bright enough to be visible in a very small telescope. Its diameter is now thought to be about 3,600 miles, so that it is appreciably larger than the planet Mercury, and seems to be the largest satellite in the Solar System. It is also the only planetary satellite to possess a relatively dense atmosphere, in which clouds may form. The atmospheric pressure at the surface of Titan seems to be about ten times as great as that at the surface of Mars.

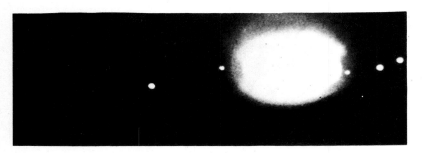

Saturn's moons
Above The inner satellites of Saturn, photographed by G. P. Kuiper: Mimas, Enceladus, Tethys, Dione and Rhea, all of which are smaller than Titan and are closer to the planet. *Left* The innermost satellite, Janus, indicated by the arrow; this was discovered by Dollfus in 1966. In both pictures Saturn itself is necessarily over exposed.

The innermost satellites of Saturn are of very low density. Little is known about Janus, which was discovered by Audouin Dollfus in 1966. It moves not far beyond the edge of the ring system, and is very elusive; it is observable only when the rings are edge-on to the Earth, a state of affairs which will not recur until 1980. Mimas, Enceladus and Tethys seem to have a density about equal to that of water, which, admittedly, is still greater than the mean density of Saturn itself. Dione, which may be almost 1,000 miles in diameter, is different. It is as dense as our Moon, and is presumably a body of the same type, whereas the inner members of the family have been described as cosmic snowballs. Rhea, too, is reasonably dense, and may be rocky. It is logical to assume that future bases will be set up on Rhea or Dione rather than upon the closer satellites.

Titan, with its planetary dimensions, is more than twice as dense as water, and has an escape velocity of 1.7 miles per second. So far as satellites are concerned, it is unique in that it has an appreciable atmosphere, and the effects of this are shown in the painting. Instead of being black, the sky from Titan is dark blue. The combination of this hue with the light from the pale, distant Sun and the glorious yellow Saturn light makes a scene which is as eerie as it is glorious. Mercury, larger and more massive than Titan and with a higher escape velocity (2.6 miles per second) is to all intents and purposes devoid of atmosphere; why, then, can Titan retain one? The reason is quite straightforward.

Titan is much colder. The surface temperature can hardly be greater than that of Saturn, which has been measured at around −240 degrees Fahrenheit. Higher temperatures mean that the particles in an atmosphere move more quickly, so that they are more easily able to escape into space. This has happened with Mercury, where the daytime heat can rise to at least 700 degrees Fahrenheit; any atmosphere which may once have existed there has long since been driven off. With Titan, the situation is different. The atmospheric particles are less agitated, and have been unable to leak away. Even so, Titan represents something of a borderline case; it has been calculated that if the temperature were raised by as little as 100 degrees Fahrenheit, the atmosphere would escape.

The existence of an atmosphere was proved in 1944 by G. P. Kuiper, by spectroscopic methods. Examining the spectrum of Titan, he detected lines and bands due to methane, the poisonous gas which is often known as fire-damp or marsh gas. There is no suggestion that an atmosphere of this sort could support life even if conditions on Titan were suitable in other ways. All the same, it is not impossible that the atmosphere may be of some use in the far future. Arthur Clarke has pointed out that methane can be used for nuclear rocket propellant, and has suggested that for this reason manned expeditions may penetrate as far as Titan before visiting the satellite system of Jupiter. Though Ganymede and Callisto, the senior members of the Jovian family, are comparable with Titan in size they do not seem to be surrounded by appreciable atmospheres. (Ganymede has an escape velocity of 1.7 miles per second; that of Callisto is decidedly lower.)

It has also been suggested that Titan's atmosphere may be of the 'cyclic' variety. Molecules escape from the relatively feeble pull of the satellite, but they are unable to break free from the powerful gravity of Saturn itself, so that they remain in orbit to be re-collected by Titan. Therefore, the atmosphere of Titan is 'recycled' regularly. On the other hand, it is also regarded as possible that the atmosphere is considerably denser than has been previously believed, in which case the recycling process may not operate. This is one of the problems which will, it is hoped, be solved by the Voyager missions.

Exploration of Titan

In the red sky is the crescent Saturn, its rings almost edge-on; the painting intentionally shows the rings tilted very slightly, in order to give a better impression of the effects of mutual reflection and shadow between the globe and the rings. An 'ice volcano' shoots fragments of frozen gases into the reddish atmosphere. Any lava would consist of liquid ammonia, methane and water.

Before manned bases on Titan, there will be robot reconnaissances; however it would be idle to pretend that they can be seriously planned as yet. A voyage to Saturn would take years if it were carried out by chemically-propelled vehicles. Even when nuclear rockets become operative, during the 1980s it will be virtually impossible to reduce the time of travel to less than some months, and once again we come back to the problem of how long an astronaut can endure conditions of reduced or zero g. Yet will there ever come a time when mankind sets up a permanently manned base in these desolate parts of the Sun's kingdom?

If the answer is 'yes', then our first remote base could well be established on Titan. The difficulties would be no greater than with the satellites of Jupiter, and the base itself would be more valuable; for one thing it could act as a relay station for more far-ranging probes, and contact with the Earth could be maintained for most of the time. During the periods when Earth and Saturn are on opposite sides of the Sun, communications would still be possible with Mars, which by then will almost certainly have full-scale colonies upon it.

From Titan the Sun seems very small. Its apparent diameter will be only 3 minutes of arc, and the amount of light and heat received will be only one-hundredth of that received on Earth. The temperature will always stay below −240 degrees Fahrenheit; in a way this will be no disadvantage, since there need be no provision against excessive daytime heat (as will be needed near the equator of our Moon).

Base on Titan

A base might well be constructed along the lines of those suitable on the Moon. Alternatively, it may be better to 'go underground', and keep only the observation rooms and the scientific equipment on the surface. Obviously there will be a need to make the colony self-supporting; to bring materials from Earth will be difficult and laborious in the extreme – and if any danger threatened the colony on Titan it would be impossible to carry out a prompt rescue operation. Neither would it be of much help to send such a rescue expedition from Mars rather than the Earth. One sometimes tends to forget the immense scale of the outer part of the Solar System; the mean distance between Mars and Saturn, for instance, is more than 740,000,000 miles.

In time it may be that all these problems will be solved. With its methane atmosphere, its dark blue sky and its icy rocks, Titan may eventually become our main outpost in the depths of the Solar System. The colonists will be scarcely able to glimpse their planet, which will be lost in the rays of the remote Sun.

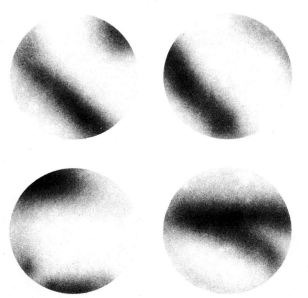

Drawings of Titan
Made by Audouin Dollfus with the 24-inch refractor at the Pic du Midi Observatory in the French Pyrenees. Light and dark areas are shown, and these are undoubtedly permanent surface features; but detail on Titan is extremely hard to make out, and can be glimpsed only with the aid of large telescopes under very favorable conditions.

The strange world of Titan *right*
Titan is a unique world, and as such is one of the principal targets for the first Voyager probe, scheduled to by-pass the system of Saturn in 1981. According to one theory there is a rocky core, surrounded by a wet, rocky mantle with water bound in the rock; outside this is a 'magma' of water containing dissolved ammonia, while the crust may be a mixture of ice or methane. Carl Sagan has suggested that there may even be 'ice volcanoes' shooting fragments of frozen gases high into the methane atmosphere covered in a layer of reddish cloud – the result of the absorption of blue and ultraviolet light. Titan is so far from Saturn that there should be no danger from any Saturnian radiation zones, particularly as these zones are certainly less lethal than those of Jupiter.

Uranus: Night Planet

Far beyond Saturn, moving at a mean distance of 1,783,000,000 miles from the Sun, moves the giant planet Uranus — just visible with the naked eye, but not known before its discovery by William Herschel in 1781. In 1984 Uranus may be bypassed by the probe Voyager 2, in which case we will have information from relatively close range. It takes 84 years to complete one journey round the Sun, and the axial rotation period is now thought to be rather less than 24 hours.

In 1977 it was established that Uranus is surrounded by a system of rings, but these are not visible directly from Earth, and are much thinner than the rings of Saturn. They are also not visible when the planet is viewed edge-on, as in the scene from the satellite Ariel.

Uranus
Above Uranus, seen from Ariel. Uranus is 119,000 miles away; the only satellite which is closer, the dwarf Miranda, appears as a crescent just above the planet. The Sun is below the horizon, to the right; its light strikes the faces of a large rock-mass. Reflected light from five fairly bright moons will render the dark portion of the planet faintly visible to eyes adjusted to the conditions, and atmospheric bands are shown.

Right Uranus and Earth compared. Uranus has 47 times the volume of the Earth, but the surface gravity is only slightly greater.

Far right The curious axial tilt of Uranus, which amounts to 98 degrees. The ring system lies in the same plane as the equator.

The strangest thing about Uranus is the tilt of its axis. The tilt is 98 degrees – and since this is more than a right angle, the rotation is technically retrograde. This means that the 'seasons' are peculiar in the extreme. First much of the northern hemisphere, then much of the southern will be in darkness for 21 Earth-years at a time, with a corresponding midnight sun in the opposite hemisphere, though for the rest of the Uranian 'year' there will be a more regular alternation of day and night.

This curious tilt will show up to advantage when we can observe Uranus from its satellites, all of which move in the plane of the equator – and this is the scene shown in the painting, in which Uranus is seen from its innermost large satellite, Ariel. We observe a pheno-menon which cannot be seen from any other satellite system: a crescent planet, the horns of which extend not from pole to pole but almost from one side of the equator to the other. The bright equa-torial band will always lie directly across the centre of the globe, but the planet will not show what we may call 'normal' phases. At the present time the planet will wax and wane between gibbous and crescent, but will never be full; by 1985, (by which time a probe should have reached Uranus), one pole will face the Sun, and the satellites will see a half-planet from any point in their orbits – the terminator lying at an angle of 8 degrees across the equator. The reason for this remarkable tilt of Uranus is not known; and neither is it shared by any other planet in the Sun's family.

From the satellites, the changing details on Uranus will be splendidly dis-played, though the surface seems to be less active and multi-coloured than that of Jupiter or Saturn.

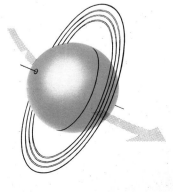

Neptune: Distant Giant

Neptune, moving at an average of 2,793,000,000 miles from the Sun, is the outermost of the giant planets. In fact, at the present time it marks the frontier of the planetary system; Pluto, which can recede to a much greater distance, is coming in toward its perihelion, and from 1979 until 1999 it will be actually closer to us than Neptune, though Pluto's inclined orbit means that it can never collide with Neptune.

Neptune was discovered in 1846. It is too faint to be seen with the naked eye, but binoculars will show it as a starlike point; it is rather bluish in colour, and it does not share the extreme axial tilt of Uranus. It is more massive than Uranus and of similar size; also in constitution the two must be very alike.

At its great distance from the Sun and from the Earth, Neptune appears small and faint to us, and surface details are very hard to make out, though there is evidence of the familiar bright and dark zones. The atmosphere consists largely of methane; at the very low temperature (around −360 degrees Fahrenheit) the ammonia is frozen out. Methane is a strong absorber of red and yellow light, which accounts for the blue color of the disk and the darkening of the limb shown in the painting – which depicts Neptune from its major satellite, Triton.

Triton is one of the largest satellites in the Solar System, and according to some estimates its diameter may be as much as 3,000 miles. In any case, Triton is larger than the Moon, and comparable with Mercury. As it also seems to be quite dense, with a reasonably high escape velocity, it would be expected to have an atmosphere, though up to now this has not been detected; we must assume that the Tritonian sky is black.

Triton moves at less than 200,000 miles from the cloud layer of Neptune – closer than the distance between the Moon and the surface of the Earth. It has a revolution period of 5¾ days, but it

travels in an east to west or retrograde direction, opposite to that in which Neptune spins on its axis. Therefore the change of view will be rapid, and the features will seem to shift quickly, as the planet itself has a reasonably fast rotation (around 22 hours).

Sunlight reaching Neptune is relatively feeble, but the human eye adapts itself readily, so that the sharp sunlight glancing over the peaks (from 'behind', and slightly to the right of the observer) seems bright in comparison with the soft, diffused radiance from Neptune itself. The high inclination of Triton's

orbit (160 degrees), combined with the planet's own normal-type axial inclination (29 degrees) takes Triton high over Neptune's northern hemisphere.

The other satellite of Neptune, Nereid, is extremely small; its diameter is less than 200 miles. It has a remarkably eccentric orbit, so that its distance from Neptune ranges between 867,000 miles and over 6,000,000 miles. It is therefore always much more remote from the planet than Triton, and it offers few advantages as an observation base. Eventually, however, an outpost may well be erected on Triton.

Neptune

Above Neptune as it appears from its major satellite, Triton. In the absence of atmosphere the sky will be black, and will be dominated by the bluish globe of Neptune, which will seem to spin quickly – partly because it really has a rapid rotation, and partly because Triton is moving round the planet in a retrograde direction. The stars will be brilliant, but the Sun will be comparatively small and feeble, and none of the other planets will be brilliant. This applies even to Uranus.

Left Neptune and Earth compared. Neptune is about the same size as Uranus, with a diameter of 31,500 miles according to a recent, improved estimate. The surface gravity 1.2 times that of the Earth.

35

Pluto: Frontier Planet

Toward the end of the last century Percival Lowell, well known for his theories of the Martian canals, studied the movements of the outermost giant planets Uranus and Neptune, and came to the conclusion that they were being perturbed by a more distant planet which had yet to be discovered. He instituted a search at Flagstaff, where he had set up an important observatory, but without success. It was only in 1930 that the planet was found. The discoverer was Clyde Tombaugh, working at the Lowell Observatory; and the planet was almost exactly where Lowell had said it would be. It was named Pluto. Since then, it has set astronomers one problem after another!

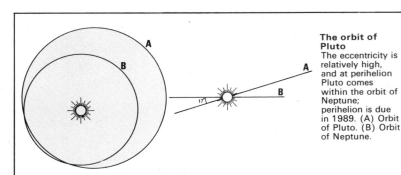

The orbit of Pluto
The eccentricity is relatively high, and at perihelion Pluto comes within the orbit of Neptune; perihelion is due in 1989. (A) Orbit of Pluto. (B) Orbit of Neptune.

Pluto is usually termed 'the frontier planet', but at present this is not true. Its mean distance from the Sun is much greater than that of Neptune, and it has a revolution period of 248½ years; but it is approaching perihelion, which will be reached in 1989 – and its orbit is relatively eccentric, so that for the next few decades it will be closer in than Neptune can ever come. However, there is no danger of a collision. Pluto's path is inclined at the relatively high angle of 17 degrees, and it cannot have even a moderately close encounter with Neptune.

Pluto escaped Lowell's personal search because it was much fainter than had been expected. Even now, when it is nearing its point of closest approach, it cannot be seen with a small telescope; and even large instruments will show it as nothing more than a dot of light. Its diameter is therefore very hard to measure, but at an early stage it was found to be

Various efforts have been made to account for this curious state of affairs. It was even suggested that Pluto might be exceptionally dense; but with a diameter of less than 4,000 miles, the required density works out at much greater than that of iron, which seems unreasonable. Alternatively, A. C. D. Crommelin proposed that Pluto is really larger than we suppose, and that what we are measuring is not the full diameter of the planet but merely a bright area reflecting the image of the Sun. This 'specular reflection' theory is not out of the question, but there are doubts as to whether Pluto can have a shiny surface; more probably the albedo or reflecting power is low.

A third theory is that Pluto is not the planet for which Lowell was looking, and that the real perturbing body remains to be found. This, too, is a possibility; but the chances of locating a planet more remote and fainter than Pluto are not very

orbit. Support for this idea is provided by the fact that Triton, alone of the large satellites in the Solar System, moves round its primary in a retrograde sense. This may have been the result of major disturbances, due perhaps to a wandering body, which allowed Pluto to leave the Neptunian system.

Pluto's light
Pluto's light is not quite steady. There are regular variations, and these seem certainly to be due to axial rotation. It is thought that the Plutonian 'day' is equal to 6 Earth-days and 9 hours. Since not even the world's largest telescope will show a perceptible disk, nothing is known of the surface features; but when Pluto is at aphelion, well over 4,000 million miles from the Sun, the temperature must be very low indeed even when compared with the −360 degrees Fahrenheit of Neptune.

Even though the escape velocity of Pluto must exceed 2 miles per second, it cannot possess an atmosphere at all similar to ours – for the simple reason that such an atmosphere would freeze. If Pluto ever possessed an atmosphere of terrestrial type, it would now lie frozen or liquefied on its surface rocks; and this is the theme of the painting, which shows the view from a cave-opening on Pluto.

We are looking out across a sea – not of water, but of methane ice. The large icicle-like structures from the roof of the cave are unchanging; on Pluto there is no movement, no sound, no life – nothing but the most utter silence and desolation. In the distance, the Sun appears as an intensely bright point, but little more; its apparent diameter is about the same as that of Jupiter as seen from Earth, though on Pluto it would still cast more light than our full moon. Stars can be seen, but there are no bright planets. From Pluto, only Neptune would appear more conspicuous than as seen from Earth, and even Neptune would be out of view for long periods.

Travel to Pluto
Long before manned craft can venture out into these depths, we ought to have close-range photographs of Pluto. Only then will we know for certain how large and how massive it really is. Spectroscopic research has indicated that much of the surface of the planet is covered with methane ice, but we have no idea whether there are any mountains or valleys there.

It is even possible that interplanetary probes might lead on to the detection of the trans-Plutonian planet, if it exists. After having by-passed Pluto, the probe should still maintain radio contact with Earth, so that its position will be known. If it is perturbed from its calculated path, the cause might be tracked down to a new planet – and once we have any clue as to the position of the hypothetical planet among the stars, a serious search can be put under way with a giant telescope.

It is much too early to speculate as

to the possibilities of establishing a base on Pluto; questions of this sort may belong not to the twenty-first or twenty-second centuries, but more plausibly to the thirty-first or thirty-second. Yet if an automatic station could be set up there the scientific information drawn from it would be of the greatest value, and there is no reason why this should not be done in the foreseeable future.

Communication with Pluto
There will be many problems to be solved, notably those of guidance and control; remember that even when an order is transmitted, it cannot be acted upon for over five hours, since it takes that time for the signal to reach the neighbourhood of Pluto. Any soft landing will have to be purely automatic, and will depend upon radar controls carried in the probe itself – as has actually been carried out already, by the Russians, with their soft-landing probes on Venus. No 'last-minute' corrections will be possible, and if anything goes wrong with the instrumentation the controllers from Earth will be helpless. Nonetheless, putting an automatic transmitting station on Pluto seems much less far-fetched now than putting a man on the Moon did at the end of the last war.

From Pluto, an observer – human or mechanical – would have admittedly a poor view of the planetary system, but there would be advantages in studying the Sun from a great distance, since very slight changes in overall brightness would be much easier to detect. Also, it is important to find out more about the amount of meteoric matter and cosmical debris in these far-away regions; and from Pluto studies could be made of those erratic wanderers, the comets, which must by-pass Pluto's orbit on their way inward to the neighbourhood of the Sun and the Earth.

Beyond Neptune and Pluto we come to a vast gulf. Even the nearest star is more than 4 light-years away, and as seen from Pluto the starry sky would look the same as it does from Earth – allowing for the lack of atmosphere. Outward, then, we look into the lonely emptiness of the Galaxy. Pluto is a world of isolation; in our painting, it is only necessary to look out from the cave-mouth at the tiny point which we know to be our splendid, blazing Sun. No lunar landscape, no Martian desert can compare with the terrifying solitude of Pluto.

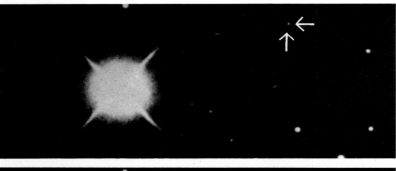

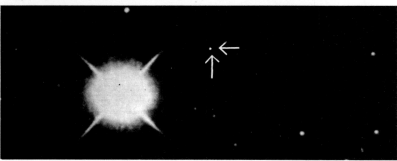

The discovery of Pluto *above*
Photographs were taken at the Lowell Observatory (Flagstaff, Arizona) by Clyde Tombaugh during his search for the planet which Lowell had predicted. The bright, very over-exposed star is Delta Geminorum,

of the third magnitude. The first photograph was taken on March 2, 1930; the second on March 5. In the interval, Pluto – indicated by the arrows – had moved perceptibly, and by just the expected amount. Within a few days Tombaugh announced his discovery.

small. Instead of being a giant, such as Neptune or Uranus, it proved to be smaller than the Earth, and probably smaller than Mars. The most recent estimate of its diameter yields a value of 3,700 miles. Of all the principal planets, only Mercury is smaller.

This presents an immediate problem. If Pluto is small, and of average density (say 5 times that of water), it must be of slight mass; it can certainly not produce observable effects upon the movements of giants such as Uranus. But it was by these very effects that Pluto was tracked down!

high, since nothing is known about its position. Of course, we may always suppose that Tombaugh's discovery was due to sheer luck – but this would indeed be a remarkable coincidence. At present the puzzle of Pluto remains unsolved.

One trouble is that we have no reliable information about its mass; it has no satellite, at least so far as we can tell. However, its diameter is not much greater than that of Triton, if the latest measurements are accepted, and it may be that Pluto is nothing more nor less than an ex-satellite of Neptune which broke free and moved away in an independent

The bleak landscape of Pluto *right*
Pluto is not a gas-giant, similar to Uranus or Neptune, but a small body, with a diameter less than that of the Earth or Mars. If it ever had an atmosphere, this must now lie frozen on the surface rocks, or perhaps liquefied; the painting shows waveless sheets of methane ice. Looking out from the mouth of the cave into the black sky we see the tiny Sun, still intensely brilliant, but too small to appear as anything but a speck with the naked eye. Beyond Pluto there is nothing but emptiness – until we come to the stars, millions of millions of miles away. Of all the worlds in the Solar System, Pluto must surely be the most inaccessible.

Comets: Wanderers in Space

Comets are the erratic wanderers of the Solar System. Unlike planets they are not solid, massive bodies; they are composed of relatively small particles, mainly icy in nature, together with extremely tenuous gas. Compared with a planet, the mass of a comet is negligible, even though some great comets of the past – such as that of 1843 – have been larger than the Sun. Due to the effects of solar wind (streams of electrified particles coming from the Sun), the tail of a comet always points away from the Sun, so that a comet which is moving outward travels tail-first.

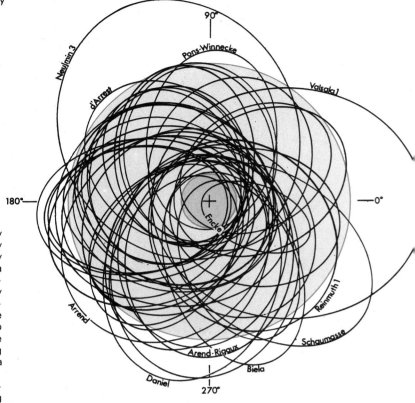

Orbits of the short-period comets *right*
Classed as belonging to Jupiter's 'comet family'. It used to be thought that comets came from outer space, and were captured by the pulls of the planets, so that they were forced into elliptical orbits, but this theory is no longer accepted.

The idea of sending a rocket probe to a comet is not nearly so far-fetched as might be thought. Indeed, the attempt may be made in the relatively near future. The only part of a comet which is of appreciable mass is the nucleus, and even this is only a few miles in diameter at most. Consequently, a probe could pass right through the head or tail of a comet without being harmed – though there is, of course, always the risk of collision with a cometary particle large enough to cause serious damage, and close-range investigations of comets will certainly be restricted to unmanned vehicles only!

Types of comet
Comets are of two main kinds. First there are the so-called short-period comets, which move round the Sun in periods of a few years; Encke's Comet, for instance, returns to perihelion every 3.3 years. Almost all these comets have very eccentric orbits, and since a comet depends upon sunlight we can observe it only when it is comparatively close to the Sun and to the Earth. Most of a comet's light is reflected, though when near perihelion the action of the Sun causes the materials in the comet to emit some light on their own account.

The short-period comets are too faint to be visible with the naked eye. Quite often it is found that the aphelion point is close to the orbit of Jupiter and it is permissible to speak of Jupiter's 'comet family'. One such comet is d'Arrest's, which has a period of 6.7 years and is due at perihelion once more in 1976. It has been under observation regularly ever since its discovery by H. d'Arrest in 1851, and its orbit is very well known. To send a probe to it, passing through the head and making measurements of the particle density – perhaps even photographing the nucleus – seems to be practicable.

Many of the short-period comets

Halley's comet *below*
Halley's Comet is the only bright comet to have a period of less than several centuries. It was visible in 1682, although its appearance had been recorded many times previously, and Edmund Halley calculated its orbit. He found that the path was almost identical with those of comets seen in 1531 and 1607, and came to the conclusion that the comets were one and the same. This series of photographs shows the development and evolution of the tail in 1910.

have no tails, and appear as dim, misty patches without structure. The only periodical comet which becomes really spectacular is Halley's, which has a period of 76 years; it last came to perihelion in 1910, and we may confidently expect it once more in 1986. Unfortunately Halley's Comet moves round the Sun in a retrograde direction – that is to say, opposite to that of the Earth and the other planets. This means that sending a probe to it will be difficult from a guidance and navigation point of view.

All brilliant comets, apart from Halley's, have periods which are so long that to all intents and purposes we may regard them as paying us one visit only. Their orbits are so highly eccentric that they are practically parabolic, and the periods may amount to thousands, tens of thousands, or even millions of years. Obviously, great comets cannot be predicted, and this also applies to the many smaller telescopic comets which have near-parabolic paths.

Great comets were fairly common in the last century; the comet of 1811 had a tail stretching right across the sky, and other brilliant visitors were seen in 1843, 1858, 1861, and 1882 (twice). In 1910 the non-periodical 'Daylight Comet' surpassed Halley's. Since then there has been a relative dearth of brilliant comets, though Bennett's, discovered by the South African amateur astronomer of that name, was conspicuous for some weeks in the spring of 1970. We cannot tell when the next great comet will appear; by the law of averages, one is considerably overdue.

Comets and meteors
Comets are associated with meteor showers, and meteoric debris is spread along cometary orbits. Each time a comet passes at its closest to the Sun, there is evaporation from the icy particles in the head; it is this which produces the tail. Constant wastage means that the short-period comets fade from apparition to apparition, and they must be regarded as short-lived members of the Solar System. Biela's Comet, for instance, was

observed to split into two portions in 1846; the twins returned on schedule in 1852, but have not been seen since. The dead comet was replaced, in 1872, by a brilliant shower of meteors. Even today, a century later, a few meteors are seen each November, marking the debris of the comet.

It used to be thought that comets came from outer space, but it is now believed that they are true members of the Solar System, even though their origin is still a matter for debate. Obviously, close-range analysis of their material would be of immense interest, and this is why plans are already being made to send a probe up with the aim of intercepting a comet.

Comet probe
In the painting, we see an unmanned probe reappearing after passing safely through a comet's head or coma. The composition of the gases has been measured, and the nucleus photographed. The high-gain antenna is turned toward the Earth, and is now transmitting information about the hydrogen envelope which surrounds the comet and stretches out for many thousands of miles. Even though this is not a Great Comet, the coma is still considerably larger than the Earth, and even at a modest relative velocity of some eight miles per second the probe has taken over an hour to pass through the comet's coma.

The tail, here fanned out by perspective, is several million miles long. It naturally points away from the Sun, which is just below the bottom right-hand corner of the picture. In the distance are seen the crescents of the Earth and the Moon; of course, these crescents also face the Sun.

Background stars can be seen shining through the comet's tail. This shows that the gases are very tenuous indeed – millions of times less dense than the atmosphere of the Earth. Though comets were dreaded in ancient times, and were thought to be the forerunners of bad news, we now know that they are harmless. Even a direct collision between the Earth and a comet would cause no more than local devastation, since there is little mass except in the small nucleus. Yet our knowledge of comets is still incomplete, and we may hope to learn more when we can make direct contact with one of these strange celestial nomads.

A comet probe *right*
An unmanned probe reappears after having passed through the coma of the comet. Spectacular though it may sometimes look, a comet has very little mass compared with that of even a small planet, and there is no reason why a probe should not be able to go right through it, though there is always a danger of collision with solid particles. The tail is even less substantial; note how the background stars can be seen through the tail virtually undimmed. The Earth and Moon are visible in the crescent form.

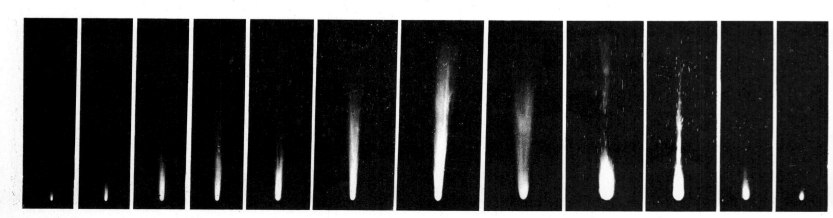

Beyond the Known Frontiers

Our rockets have taken men to the Moon. Within the next few decades the first explorers will land on Mars, and unmanned probes will voyage out to the cold depths of the Solar System. Yet we must not fall into the trap of supposing that we are 'conquering the universe'. All we are doing is sending our messengers to our own local region. Beyond lie the stars – so far away that to reach them will need new techniques at the moment beyond our understanding.

The Sun is a star. It is no larger, hotter or more luminous than many of the stars visible on any clear night; indeed, astronomers rank it as a dwarf. Although it is of fundamental importance to ourselves, since we belong to its system, it is of practically no importance in the Galaxy considered as a whole. Even the nearest star beyond the Sun is more than 24 million million miles away. Before we can seriously discuss interstellar travel, we must appreciate how vast the Galaxy really is.

Our Sun, a normal star, is over 5,000 million years old. It is attended by a family of planets, of which one – the Earth – is inhabited. There is no reason to doubt that other stars also have planetary systems circling them, and it is only logical to suppose that many of these planets must support life. Unfortunately, planets of other stars cannot be observed directly by means of our present-day equipment. A planet is much smaller than an ordinary star, and has no light of its own; it shines only by reflection. No telescope we have yet built or planned would be capable of showing even a large planet moving around a comparatively close star. We have excellent evidence that such planets exist, but this evidence is, as yet, indirect.

Certainly there are plenty of stars from which to choose. The Galaxy contains approximately 100,000 million suns, of all sizes, luminosities and stages of evolution. Some are immensely powerful, so that they would outshine the Sun just as a searchlight outshines a glow-worm; others are feeble. Some are so huge that they could contain the whole path of the Earth round the Sun, while others are smaller than the Moon. Some are millions of times less dense than the air we breathe, while others are so 'heavy' that many tons of their material could be packed into a space the size of a letter 'o' on this page. And though most of the

The most we could hope to do would be to identify signals sufficiently rhythmical to be classed as non-natural. The distances involved would preclude any exchange of information, even if a suitable code could be worked out. If signals were picked up from, say, a star 10 light-years away, then the signals would already be 10 years old by the time they reached us – and any reply would not arrive at the distant system for a further 10 years. This is an extreme case; there are not many stars as close as this.

The necessary speed
According to modern theory, no material body can move at the velocity of light; but there is no theoretical objection to reaching a speed very close to it. An atomic rocket, able to accelerate over a very long period, may one day be built for interstellar flight. It sounds fantastic as yet – but much less fantastic than reaching Mars would have seemed to Julius Cæsar! Alternatively, some method may be found which does not involve material transfer. It is too early even to speculate about problems of this order, but they may well be tackled during the centuries ahead.

The Galaxy would appear as a flattened system if it could be observed edge-on. The overall diameter is 100,000 light-years, and the shape is roughly as

It is thought that a star begins its career by condensing out of the thinly spread interstellar material, and that nebulæ may be regarded as stellar birthplaces.

In the spiral arms of the Galaxy there is considerable interstellar matter, and many nebulæ. The most brilliant stars are extremely hot, and white or bluish-white. Objects of this sort are known as Population I, a term established originally by the late Walter Baade. Elsewhere – for instance, in some regions near the galactic center – there is much less material available to form fresh stars, and the most conspicuous objects are large, powerful red stars. These objects are classed as Population II.

How a star develops
With this information, we can begin to examine the way in which a star develops. As we have seen, it begins by condensing out of nebular material. At first it does not shine; but as it contracts, under the influence of gravity, its inner regions become hot. When the temperature has risen sufficiently, nuclear reactions begin. The star settles down to a long period of stable radiation; it is said to belong to the Main Sequence. This is the present condition of the Sun.

The production of energy
Stars are not 'burning' in the usual sense of the term. Inside the Sun, nuclei

one. At distances of millions of light-years we can see others, many of which are spiral in form. The famous Andromeda Spiral, which is dimly visible to the naked eye, is a spiral larger than ours, and so far away that its light takes 2,200,000 years to reach us. Yet even this system belongs to what we call the Local Group – a collection of over twenty galaxies ranging in size and brightness from great spirals like M31, to faint dwarfs barely detectable even with the largest telescopes.

No telescope can show a star as anything but a point of light, and our information comes chiefly from instruments based on the principle of the spectroscope. Starlight is split up, and we can find out what elements are present in the stars themselves. We can also discover whether the stars are approaching us or are moving away. The

The Milky Way Galaxy
Below left The Milky Way Galaxy, as it would be seen edge-on. It is a flattened system, 100,000 light-years in diameter; the Sun lies 32,000 light-years from the hub. Surrounding the main system is the 'galactic halo', made up of compact globular clusters as well as isolated stars. *Below* The Galaxy as it would be seen in plan. In this view the spiral arms are shown; the Sun is situated near the edge of one of them. As yet we cannot pretend that we know why spiral arms form, or whether they are permanent features throughout most of the evolutionary cycle of a galaxy.

stars shine steadily for year after year and century after century, some are variable in light, and a few are violently explosive.

Contact across the Galaxy
How are we to get in touch with other civilizations, far away across the Galaxy? We cannot hope for direct contact by means of rocket probes, piloted or automatic. Represent the Earth–Sun distance by one inch, and the nearest star will be four miles away. Moving at velocities we can now attain, a rocket would be thousands of centuries on the journey. The only thing which seems to move quickly enough is light – or, to be more precise, any form of electromagnetic radiation. It is not impossible that we might be able to pick up artificial transmissions from another system, and efforts to do so have already been made, though without success.

shown in the diagram at the left. The Sun, with its system of planets, lies close to the main plane, but well away from the center; the distance to the galactic hub is about 32,000 light-years.

Viewed in plan, it would become obvious that the Galaxy is spiral, not unlike a huge pin-wheel; the Sun is near the edge of one spiral arm. The Galaxy is rotating, and the Sun takes approximately 225,000,000 years to complete one journey round the center. This period is sometimes known unofficially as the cosmic year. One cosmic year ago, the most advanced creatures on Earth were amphibians; even the fearsome dinosaurs were yet to make their entry on the terrestrial scene.

The Galaxy contains many features besides individual stars. There are, for instance, clouds of dust and gas which are known as nebulæ; the most famous example is in the constellation of Orion.

of hydrogen atoms are combining to form nuclei of the second lightest element, helium. Each time a helium nucleus is formed from hydrogen, a little energy is released and a little mass is lost. It is this energy which keeps the Sun shining; the mass-loss amounts to 4 million tons per second, but although this seems very rapid we may be confident that the Sun will continue radiating steadily for at least 6,000 million years in the future. Eventually, of course, its supply of available hydrogen will begin to fail, and the Sun will change radically. It will become a red giant, and for a period will be much more luminous than it is today, though its surface will be cooler.

We cannot expect life on Earth to survive the Sun's change into a red giant; but if humanity still exists, no doubt some way will be found to avoid destruction.

Our Galaxy is by no means the only

spectrum of a galaxy is made of the combined spectra of its millions of stars, but it can be interpreted; and it has been found that all the galaxies, apart from those in the Local Group, are racing away from us. It is believed by many astronomers that some enigmatical objects known as quasars, which appear smaller than galaxies but may be essentially similar to them, are even more remote and receding even more rapidly.

This does not imply that we are in any special position. The entire universe is expanding. Whether this will continue indefinitely we cannot tell, neither do we know anything about the way in which the universe was created. All we can do is to study its evolution; we are totally ignorant of the fundamental 'beginning'.

From Earth we obviously cannot see our Galaxy in its entirety. The center of the system lies in the direction of the wonderful star-clouds in the constella-

tion of Sagittarius (the Archer), but our view of the center itself is cut off; there is too much interstellar material between it and ourselves, so that light waves are absorbed. Longer wavelength radiations, known as radio waves, can however come through, and are collected and analyzed by radio telescopes such as the famous paraboloid at Jodrell Bank. On the other hand, we can have overall views of many other spirals, and so we can form a very good idea of what our Galaxy would look like if it could be seen from beyond.

In the painting, it is night on an inhabited planet of a 'stray' star moving some 200,000 light-years outside the Galaxy. The spiral nature of the arms, glowing with hot blue Population I stars, is clearly shown, as is the central hub, dominated by the cooler but very luminous red giants characteristic of Population II. Individual stars cannot be

seen at this distance; only the star-clouds are visible, and between the arms are dark patches – dust clouds which are not lit up by stars, and which may be called dark nebulæ. Well above the horizon, to the left in the painting, is the solitary moon of our hypothetical planet, casting its radiance on to the landscape below.

Life on such a planet

What beings could live here? This is a question we cannot yet answer. If a planet has a suitably even temperature, if there is atmosphere and water, it is reasonable to suppose that life will develop; and if conditions are similar to those on Earth, intelligent beings could well be similar to ourselves. If the environment is different, then the life-forms will be different, though they may be equally advanced – or even more so.

On this planet of a stray star, one

feature will be missing from the night sky: the Milky Way. To us on Earth, the Milky Way is a glorious band of light stretching from one horizon to the other. It is made up of stars, together with cosmic clouds both bright and dark – but the stars in it are not crowded together, as they appear. The Milky Way is mainly an effect of perspective. When we look along the main plane of the Galaxy we see many stars in much the same direction, and it is this which produces the band of radiance. On a world outside the Galaxy nothing comparable would be seen; part of the sky would be almost starless, but in compensation there would be the glorious spiral shown here.

Through a telescope, individual stars could be seen. One of them, an insignificant yellow dwarf, is here just above the planet's moon – and indeed lost in its glare. Outwardly it has nothing to single

it out for special attention. How could alien beings tell that this obscure star is the center of a planetary system, and that the inhabitants of one of those planets have already taken their first steps into space? In fact, their view of the Sun would be 200,000 years out of date, and must go back to a time before civilization on Earth had begun.

We can hardly doubt that many other-world astronomers are at this moment looking at the Sun from their observatories built upon planets of other stars. And one day, contact may be made.

Interstellar Communication

Communication with astronauts on the Moon is a very easy matter indeed. Neither will there be any difficulty with Mars, or indeed with any of the worlds in the Solar System. The slight delay (up to five hours or so with the outermost planets) will be no more than irritating. However, communicating with hypothetical civilizations living on planets moving round other stars is a problem of a different order. Certainly we can have no hope of sending a twentieth-century type rocket probe there; the time of travel would be hopelessly long. Even light, moving at 186,000 miles per second, takes more than four years to reach us from the nearest star. Radio waves, of course, move at the same velocity as light, and if we are to establish contact it can only be by means of radio. But apart from all other difficulties communication will be slow. Even with Proxima Centauri, a message transmitted in (say) 1980 would not arrive until 1984, and no reply would be possible before 1988 at the earliest.

All the available evidence indicates that planet families are common in the Galaxy. If this is so, then we may assume that there are also many civilizations, some of which may have achieved a far higher level of technology than we have as yet managed to do. They, too, must have their radio telescopes and their radio astronomers; what are our chances of 'tuning in' to them?

Obviously, the difficulties are very great. It is not easy even to pick up natural radio emissions from ordinary stars; in fact very few stars have so far been identified in the radio range. Early in 1972 C. M. Wade and R. M. Hjellming, at Green Bank, discovered that both Algol and Beta Lyræ are radio sources; the only other stars of the same kind are Antares B and the star associated with the X-ray source Cygnus X-1. (There is a possibility that the invisible component is a Black Hole.) Other radio sources are supernova remnants, gas-clouds, galaxies and quasars. Artificial

Haystack Radio Telescope *right*
The Haystack radio telescope of the Massachusetts Institute of Technology, Lincoln Laboratory. Above is a drawing showing some details of the radio telescope, which has a diameter of 120 feet. An operator is shown adjusting the 9-foot secondary reflector (A). A 2-ton special purpose electronics box (B), is hoisted into place to be plugged into the antenna. The boxes are interchangeable to permit different kinds of experiment. The covering 'radome' is 150 feet in diameter, and constructed on geodetic principles.

transmissions from a planet moving round another star would inevitably be feeble and hard to identify.

Project Ozma

In 1960 radio astonomers at Green Bank, West Virginia, began an ambitious programme known officially as Project Ozma. With powerful equipment, they concentrated upon the two nearest stars which are reasonably like the Sun — Tau Ceti and Epsilon Eridani, both of which are rather smaller and cooler than the Sun and are more than 10 light-years away. It was thought possible that either star might be attended by an inhabited planet, and the Ozma researchers were trying to pick up rhythmical signals which might be interpreted as artificial. They selected a wavelength of 21.2 centimetres, since this is the wavelength of the radio signals emitted by the clouds of cold hydrogen spread through the Galaxy. It was logical to believe that other radio astronomers, wherever they might be, would also be devoting their attention to this particular wavelength.

Ozma was (not unexpectedly!) negative, and was soon discontinued; but it was not unreasonable, and in the future it may be tried again. A suitable site for the equipment would be on the far side of the Moon, where the powerful transmitting stations on Earth can cause no interference with incoming signals. Admittedly the chances of success are slight; but they are not nil, and at present they represent our best hope of proving, once and for all, that other civilizations exist.

Radio contact

Conversely, it may well be that alien radio astronomers are at this moment 'listening out' in the region of the Sun. They may know nothing of the Earth, which is so small that across galactic distances it will be excessively hard to detect; but they may pick up our radio signals. If so, they will know that 'beings' of some kind exist in the Sun's system. Clearly, radio methods of contact between one planetary system and another are restricted, and information will always be out of date. If, to take an extreme example, we could pick up a rhythmical signal from a planet moving round a star 600 light-years away, all we could prove is that radio astronomers existed there 600 years ago! Perhaps, in the future, other methods such as thought-communication will be developed; but as yet we have no idea of how this might be done, so that speculation is both endless and pointless.

Horn antenna at Andover, Maine *left*
Situated inside a 'radome', one of the first major uses of this radio telescope was tracking the Telstar satellite in 1962. The size of the antenna is indicated by the man standing beneath the 'dish'.

Interstellar contact *right*
An alien radio telescope, directed upward at an alien sky. The design is much the same as that of our own Jodrell Bank paraboloid; but in the sky our hypothetical planet has a vast moon, with its sun low down over the horizon. Perhaps the scientists there are studying the region of the Sun, many light-years away; perhaps they will be successful in establishing that a peopled planet exists in the Sun's system — and perhaps, one day, they will be able to contact us.

First Star Beyond the Sun

The nearest of all the bright stars is Alpha Centauri, which is never visible from Europe, but is a brilliant object in southern skies; of all the stars only Sirius and Canopus outshine it. It is made up of two components, moving round their common centre of gravity and making up what is termed a binary system. The distance of the Alpha Centauri pair is 4.3 light-years. (One light-year is equal to 5,880,000 million miles.)

There is a third member of the Alpha Centauri system. This is Proxima, which is much fainter than its luminous companions; although it is the nearest star to the Earth, approximately one-tenth of a light-year closer than Alpha Centauri, it is far too dim to be seen without a telescope. It is a celestial glow-worm – a feeble red dwarf.

Proxima Centauri is a star of very different type from the Sun. Its surface temperature is much lower, and its total luminosity is only 0.0001 of that of the Sun. Its evolutionary career, too, must have been different, simply because of its lesser mass. If our Sun were as feeble as Proxima, a planet moving at a distance of 93,000,000 miles would be a chilly world indeed – much too cold to support life of our kind. Therefore, if Proxima has a habitable planet, we must look for it in a region much closer to the red surface.

Perhaps it is misleading to say 'look for it', because even if Proxima were attended by a planet as large as Jupiter we would have no hope of seeing it with our present-day telescopes. Fortunately, there are other methods. Relatively close stars have perceptible individual or proper motions, and shift slowly against the background of more distant stars; Proxima does so at the rate of 3.75 seconds of arc per year. This means that it takes 600 years to move by an amount equal to the apparent diameter of the Moon as seen from Earth. Slow though this motion may seem, it is easily measurable from year to year.

Barnard's Star

Another faint red dwarf, Barnard's Star, has an even greater proper motion (over 10 seconds of arc per year) even though it is more remote than Proxima. Barnard's Star is not, however, moving smoothly. It is 'weaving' its way along, and it has been found that this irregular motion is caused by the presence of an invisible body which moves round the star and pulls it out of position. The mass of the companion is too small for it to be a star, and so it is presumably a planet; indeed, there may even be two planets 'accompanying' Barnard's Star. This is a reliable indication that the retinue of planets moving round our Sun is very far from being unique.

There have been suspicions that Proxima, too, may have a planet moving round it. The evidence at present is slender, but the planet may exist. The painting shows the view which might be expected from it. The planet is relatively near its weak red sun, around which (it is calculated) it orbits in 10–12 days; the limb of Proxima is not sharp, but is clearly diffuse. The landscape is completely hypothetical; we see eroded black, basaltic rocks, as lonely and desolate as anything on our Moon, though the planet retains a thin atmosphere which is replenished by occasional feeble bursts of gas from volcanic vents. No sedimentary rocks hint at past life, though water survives in a lake fringed by glittering ice-crystals. In the sky a solitary moon is silhouetted in transit against the dull redness of Proxima. The two bright stars of Alpha Centauri are 'behind' us, and at the moment add no perceptible illumination, though they will be brilliant during the planet's period of night. Perhaps this scene is duplicated many times on planets of other red dwarf stars in the Galaxy.

Planet of Proxima Centauri *above*
The scene from a hypothetical planet orbiting Proxima Centauri. The landscape is desolate; in the dark sky the stars shine down, with the constellation patterns very similar to those we know. To the right of the black disk betraying the transit of a moon we see the W of Cassiopeia, but there is an extra star; this is the Sun, which from Proxima will be conspicuous but not glaringly so. It will convert the W into a constellation which inhabitants or future interstellar travellers may well nickname 'the Switchback'. The Southern Cross, below the horizon in this painting, will lack one of its two 'pointers' – Alpha Centauri itself, which will appear as a pair of distant suns casting light on to the bleak rocks of the orbiting planet.

Red Supergiants

Of all the stars the huge red giants are among the most remarkable. It used to be thought that they were comparatively young, but we now know that they are old; they have used up their hydrogen 'fuel', and are shining because heavy elements are being synthesized inside them. An extreme example – such as Zeta Aurigæ, shown in the painting – may have a diameter hundreds of times greater than that of the Sun; indeed, Zeta Aurigæ could contain the whole of the Solar System out as far as the asteroid belt. If red giants evolve from Main Sequence stars (and of this there is virtually no doubt) they will swallow up their inner planets; surviving planets must be orbiting at a respectful distance from their swollen sun. From being cold worlds, they must have become searing hot and the prospects for the survival of any life-forms on them do not seem to be high. With Zeta Aurigæ the situation is complicated by a binary companion: a very hot bluish star much more powerful than the Sun.

There are no giant or supergiant stars anywhere near the Sun. All our close stellar neighbours are dwarfs – even Sirius in the Great Dog, which looks so magnificent in our skies, is a humble Main Sequence star. Therefore, even if and when interstellar flight becomes possible, it will not be for a long time that we can investigate the planetary systems of giants – assuming that such systems exist.

Planets of giant stars are likely to be uncomfortable places so far as climate is concerned. Even if the central sun is a single star rather than a binary such as Zeta Aurigæ, it is quite likely to be variable. Many of the red giants and

Main Sequence. This is the present condition of the Sun, where the temperature near the core is of the order of 14 to 15 million degrees centigrade. Hydrogen is being steadily converted into helium, and the output is steady – discounting very minor variations over a long period, causing the changes in temperature of the Ice Ages which have affected our world at intervals throughout geological history.

The state of balance

After thousands of millions of years there is no more available hydrogen in the core. Up to this time the star has been in a state of balance between gravity,

has left the Main Sequence, and has moved off into the giant branch – to the upper right in the diagram below. The yellow, white or bluish sun has become a red giant.

The Hertzsprung-Russell diagram

The diagram given here is of great importance to theoretical astrophysicists. It is known as a Hertzsprung-Russell or H-R Diagram, in honour of the two astronomers who drew it up more than half a century ago. Originally it was thought that a star 'slipped down' the Main Sequence from the top left to bottom right of the diagram, shrinking steadily and becoming weaker; but we

Even if the Earth survives, it will become much too hot to retain any atmosphere or surface water, so that life here cannot continue indefinitely.

Zeta Aurigæ, shown in the painting, has already reached the red giant stage; any inner planets will have been destroyed. Only the more remote planets in such a system can still exist, and even these are likely to be uninhabitable. When the first explorers reach them, the scene will be one of total sterility, but also one of weird beauty; with Zeta Aurigæ, the combination of the red super giant and the hot bluish companion would produce colour effects which defy description.

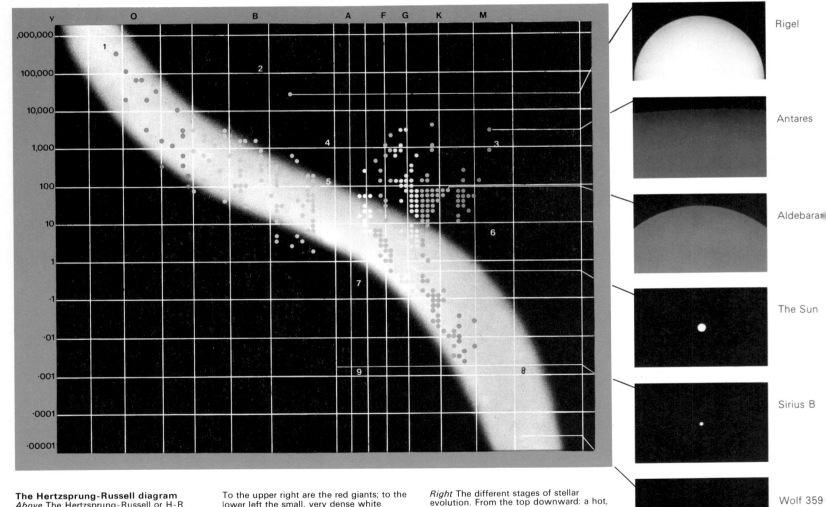

The Hertzsprung-Russell diagram
Above The Hertzsprung-Russell or H-R Diagram. The stars are plotted according to their spectral types and surface temperatures (X) and their absolute magnitudes or luminosities (Y) in terms of the Sun. Most stars lie on the band of the Main Sequence.

To the upper right are the red giants; to the lower left the small, very dense white dwarfs, which have exhausted their nuclear energy. After its period on the Main Sequence, a star such as the Sun becomes a red giant and then collapses into a white dwarf.

Right The different stages of stellar evolution. From the top downward: a hot, massive and very luminous star; a red supergiant; a less extreme red giant; a Main Sequence star; a white dwarf, and a feeble red dwarf. In order: Rigel, Antares, Aldebaran, the Sun, Sirius B, Wolf 359.

Rigel

Antares

Aldebaran

The Sun

Sirius B

Wolf 359

supergiants fluctuate considerably in their radiation output over relatively short periods of a few weeks or months. This variation is intrinsic, and is not due to an eclipsing companion as with Algol or Beta Lyræ – even though it is possible that some red giants may also have companions.

As we have seen, a star begins its luminous career by condensing out of interstellar material. When its inner temperature is sufficiently high, nuclear reactions begin, and the star joins the

tending to pull all the material to the centre, and radiation and gas pressure, tending to distend the globe. When the production of energy from the hydrogen-into-helium process is halted, the star begins to collapse, but the process does not get very far. As the core temperature rises still further, the helium which has been accumulated there begins to undergo reactions in its turn. These are succeeded by yet others; meanwhile the outer layers of the star have expanded, cooling and changing colour. The star

now know that this is wrong. Red giants are older than Main Sequence stars, or are at least more advanced in their cycle of evolution.

The Sun

A star such as the Sun will go through its red giant stage after leaving the Main Sequence. The Sun may become at least 100 times as luminous as it is now, though its surface temperature will be lower; it will swell out, engulfing Mercury, Venus and probably the Earth.

Planet of Zeta Aurigæ *right*
Intriguing bi-coloured shadows on an imaginary planet of the eclipsing binary Zeta Aurigæ. This fantastic system consists of a vast red supergiant over 200,000,000 miles in diameter, together with a hot bluish-white star with a diameter of 3,000,000 miles; this hot star is much more powerful than the Sun. In the painting, the blue star is about to be eclipsed; its light is already dimmed as it passes through the tenuous envelope of gas surrounding the supergiant. The surface of the red star is obscured by an interlacing network of glowing prominences. The hypothetical planet in the painting is 700,000,000 miles from the red supergiant.

47

Twin Suns

Though there must surely be millions of 'other Earths' in the Galaxy, not all stars are suited to be the centers of planetary systems. The most likely candidates are stable, single stars such as the Sun. A planet moving round a binary star would have a strange, uncomfortable climate – unless the two suns were close together and luminous enough to warm a planet circling at a comparatively great distance.

Binary systems of all kinds are known. Sometimes the component members of the pair are very widely separated, so that the mutual revolution period amounts to thousands or even millions of years. In other cases the two suns are almost in contact, so that each is drawn out into the shape of an egg. From Earth we cannot see the individual members of such a pair, but spectroscopic observations can give us a surprising amount of information. Beta Lyræ, in the constellation of the Harp, is one such binary, but many others of the same type are known.

In 1667 an Italian astronomer named Montanari noticed that there was something very strange about Algol, a fairly bright star in the constellation of Perseus. Stars are graded according to their magnitudes, relating to their apparent brightness (not their real luminosity); very bright stars are of magnitude 1, while the faintest stars normally visible to the naked eye are of magnitude 6. Algol is normally of the second magnitude, roughly equal to the Pole Star. Montanari found that every two and a half days it gave a very slow 'wink', taking some hours to fade down to below the third magnitude and staying at minimum for twenty minutes before regaining its lost luster. By what seems to have been sheer coincidence, the Arab astronomers of a thousand years ago knew the star we call Algol as 'the Demon Star'.

the fainter star and ourselves.

Today many of these eclipsing binaries are known. They are extremely valuable to astronomers, because they can provide information about the sizes and masses of the individual stars in the binary. With a binary system in which there is no eclipse, we can find out only the combined masses of the two components. Of course, there is no real difference between an eclipsing and a noneclipsing binary; it is merely a question of the angle at which the orbit is tilted relative to the Earth.

Beta Lyræ
The second eclipsing binary to be discovered was Beta Lyræ, sometimes still known by its old proper name of Sheliak. It is easy to find, since it lies

hot and bluish-white. The secondary is less hot, but its spectrum has not been observed, so that we are by no means certain what it is like. It has always been taken to be a star with a surface rather hotter than that of the Sun. Recently there have been suggestions that it may be an altogether different type of object – a 'black hole', or collapsed star; but for the moment we may regard Beta Lyræ as a normal eclipsing binary.

The scene from a hypothetical planet orbiting such a pair is shown in the painting. It is an incredible spectacle. The two suns are so close together that they almost touch, and an observer on the planet would see their distorted shapes. Moreover, the stars are surrounded by streamers of gas. Glowing hydrogen is ejected from the equator of the larger star, which is

Planet of Beta Lyræ *right*
The scene from a hypothetical planet orbiting Beta Lyræ; we are looking out from a crevasse on the surface. The sky is dominated by the twin stars and the spiral of expanding gas.

the two suns, almost or quite in contact, would act as one mass, but there would be wide variations in surface temperature if the plane of the orbit were such that the suns mutually eclipsed each other at each revolution. It would certainly be unwise to claim that no life could exist on a planet of Beta Lyræ, but it is logical to say that the chances are lower than with a single star of solar type. On the other hand, there is no reason why planets in the system could not be visited – once the problems of interstellar travel have been solved!

The view from the planet
Beta Lyræ may be an exceptional object, but there are many other binaries where the view from an orbiting planet would be very much as shown in the painting. Beyond the twins, with their streamers of gas, would lie the other stars of the Galaxy, but this time the star patterns would be unfamiliar to anyone used to the sky of Earth. The situation is not the same as from Proxima, which is relatively so close to us that the overall view would be much the same as ours. Beta Lyræ, remember, is over 1,000 light-years away, so that everything would be changed. The Sun would be a very dim object much too faint to be seen with the naked eye by an observer with sight equal to ours.

Because Beta Lyræ is a much less stable system than that of Alpha Centauri (and, of course, far less stable than the Sun), its evolution must be quicker, and neither does it have so long an expectation of 'life' before drastic changes occur within it. If there is a life-bearing planet, alterations in the twin suns will certainly destroy this life long before Earth becomes uninhabitable.

With a Beta Lyræ-type binary in which the components are more widely separated, the chances of finding a planet become less, because the two stars would not act as a single controlling mass, and the orbit of the planet would be erratic, resulting in wild fluctuations of temperature. Sometimes one sun would be close, sometimes the other; then there would be periods when both stars shone together, scorching the surface of the planet in a blaze of heat. Yet there must be worlds like this across the Galaxy; and in the distant future it may be that some of them will be explored by men who have traveled hundreds or thousands of light-years across the depths of interstellar space.

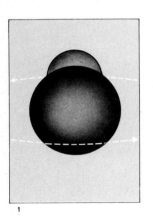 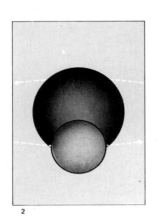  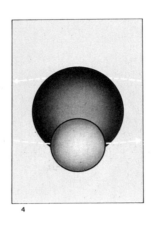

1   2   3   4

Diagram of an eclipsing binary *above*
The eclipsing binary Algol. In the first and third positions the smaller but more luminous star is partly hidden by the larger, cooler component, and from Earth we see the principal eclipse; from the second magnitude Algol declines to below the third. In the second and fourth positions the faint component is partly hidden by the more brilliant. The result is a slight decrease in the light we receive; this secondary minimum is detectable by means of sensitive instruments known as photometers.

Montanari did not know the cause of this strange behavior, but Algol's fluctuations were explained in 1783 by John Goodricke, a young deaf-mute astronomer. Algol is not genuinely variable in light. It is a binary; one component is larger but less brilliant than the other. Every two and a half days the fainter component partially eclipses or hides the brighter, as shown in the series of diagrams above, so that the total light we receive from the system drops. There is a much less obvious drop in the light when the brighter component passes between

close to the brilliant blue star Vega, which is almost overhead as seen from Britain and the northern United States during summer evenings. In fact Beta Lyræ is much the more distant of the two. Vega is a mere 27 light-years from us, while the distance of Beta Lyræ is about 1,100 light-years; we see it as it was in the reign of Alfred the Great.

Like Algol, Beta Lyræ has two main components, much too close together to be seen individually. The two are less unequal than in the case of Algol, and so the behavior is different. When we construct a light-curve, plotting time against the changing magnitude, we find the effect shown in the diagram. Variations are always going on; there are alternative deep and shallow minima. During a deep minimum the more luminous component is partly hidden by the fainter; at a shallow minimum it is the fainter star which is covered up. The revolution period is 12 days 22 hours 22 minutes, and this appears to be increasing at ten seconds per year.

The larger component of Beta Lyræ is

spinning round rapidly. Some of the material is 'caught' by the fainter, smaller star; there is a constant loss of material into space, so that the overall effect is that of an expanding spiral. We are observing from a deep crevasse on our planet; the shadows and the color effects are ever-changing.

The planet's climate
What kind of climate could be expected on a planet moving round a system like this, and what sort of life could conceivably evolve there? We must assume that the planet has an atmosphere if there is to be any life, but the hotter star of Beta Lyræ radiates strongly in the short-wave end of the spectrum, and some of these radiations would be lethal to terrestrial-type life were they not effectively screened. The planet would have to be far enough from the central pair to avoid the effects of the gas streaming, even though the ejected material is certainly very rarefied. The orbit of the planet would be fairly stable, since from a gravitational point of view

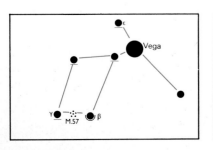

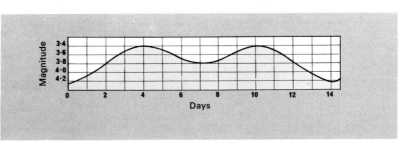

Beta Lyræ
Left Light-curve of Beta Lyræ. The period is just over twelve days, and includes two maxima; the minima are alternately deep and shallow, as the two members of the binary are much less unequal than with Algol. M.57 is the Ring Nebula, seen in a photograph on p. 56.

Far left Position of Beta Lyræ. It lies near the brilliant Vega, and forms a pair with the third-magnitude star Gamma Lyræ. At its brightest, Beta is slightly inferior to Gamma.

Triple Suns

Binary pairs are of various types; sometimes the two components are almost touching, as with Beta Lyræ, while in other cases the members of the pair are a very long way away from each other. As well as physical doubles, the Galaxy also contains triple and multiple stars; Zeta Cancri, shown in the painting, is a splendid example of a triple system with components of contrasting colors. The two close stars are wider apart than in the case of Beta Lyræ, but tidal distortion still pulls out each member into an egg-shaped form.

Seen from the Earth, Zeta Cancri appears as a rather obscure star in the ill-defined constellation of the Crab. It is of the fifth magnitude, so that it is distinctly visible with the naked eye on a clear night. Through a reasonably powerful telescope it is seen to consist of two stars, one orange and one white, which make up a binary pair with a revolution period of 59.6 years. Further away, and easily detectable in a small telescope, is a yellow star, orbiting the primary pair in a period of 1,150 years. The paths of the three members of the system are decidedly complicated, as shown in the diagram, so that a planet will have a variable and erratic orbit, with wide ranges in temperature and illumination.

The scene shown in the painting is a faithful representation of what would be observed from a planet of the Zeta Cancri system. Here, even more than with Beta Lyræ, we have spectacular lighting effects, because the three components are quite similar in luminosity.

since been driven off, so that the sky is star-studded and black even though the two suns loom so large. The distant member appears as a small sun, and is surrounded by a swarm of cosmic dust, recalling our own Sun's Zodiacal Light. Because of the long revolution period, the distant companion would seem to move very slowly against the starry background. When the planet moves between it and the close pair, there can be no true darkness; 'night' will occur for only part of the planet's 'year'.

Evolution of the system
The orange component of Zeta Cancri has already left the Main Sequence, and moved off to the upper right part of the H-R Diagram into the giant branch. Here, too, we must therefore expect that the planetary system will have suffered severely, with the inner members being destroyed and the formerly chill outer planets becoming suddenly super tropical. Interstellar travelers will indeed

are plenty in the Galaxy. Much the most famous is that of the Pleiades or Seven Sisters, which is very prominently visible with the naked eye; it lies in the constellation of Taurus (the Bull), not far from the red giant Aldebaran.

The Pleiades cluster contains at least seven stars visible without optical aid (keen-sighted people can see more than a dozen), and the total number of stars exceeds 200. The distance of the cluster is 410 light-years; the leading stars, headed by the third-magnitude Alcyone, are hot and bluish-white. Mixed in with the cluster-stars is a gaseous nebula, which shines by reflection and is well shown only in photographs taken with large telescopes.

Many other open clusters are known, some of which are visible with the naked eye; for instance we have the Hyades round Aldebaran (though Aldebaran is not itself a member of the cluster), Præsepe in Cancer, and the lovely 'Jewel Box' round Kappa Crucis in the Southern Cross. Yet even inside these clusters, the individual stars are still widely separated – by many thousands of millions of miles – and collisions can hardly ever occur. It has been calculated that an ordinary star has very little risk of suffering a collision, or even a close encounter, with another star during the whole of its lifetime.

Yet if it were possible to go to a planet inside an open cluster, the night sky would be glorious, with many stars brighter than Sirius appears to us. Some of them would cast perceptible shadows, and darkness would be unknown.

The fear of darkness
How would this affect any life-forms on such a planet? Of course we can only speculate; but we can imagine that darkness would be their greatest fear, simply because it would not occur except in places closed off from the sky. In the daytime there would be the glare of the local sun – perhaps a bluish-white star, though clusters also contain red giants and feeble dwarfs. After the sun had set, there would be the hundreds of thousands of night-time suns, still casting a brilliant light across the surface of the planet.

No cluster lies near us in the Galaxy, so that before we can send astronauts to such a place we must learn how to travel hundreds of light-years. As yet we have no idea of how it might be done; but the prospect is not as fantastic as it may sound. Once the method has been found, it may be a relatively minor step between traveling four light-years and traveling four hundred. Soon after we have sent our pioneers out to Proxima and Alpha Centauri, we may also be able to send them to the vividly-colored system of Zeta Cancri or to the inner part of the cluster of the Pleiades. Our Sun will have faded into the distance; but there will be suns in plenty to take its place.

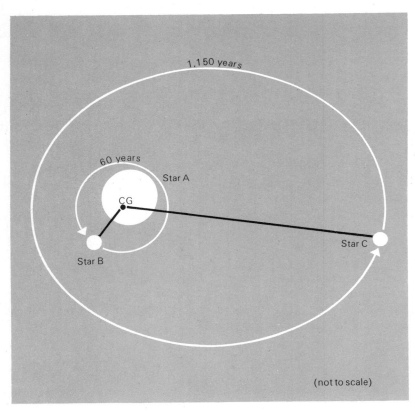

1,150 years

60 years

Star A

CG

Star B

Star C

(not to scale)

The smaller member of the close pair is white, but in the painting it appears bluish, because of the contrast effects in the system. A 'bridge' of tenuous material is possible, between the white star and the orange giant, which is much larger than the Sun even though not nearly so extreme an example of a giant star as the red component of Zeta Aurigæ.

Lifeless planet
A planet in a system of this kind is likely to be lifeless; here we see a volcanic-type landscape, with a trace of activity remaining in the vent to the left of the painting. Lava flows cover the barren surface, and any atmosphere has long

find very strange and beautiful sights awaiting them.

Triple systems are spectacular by any standards, but in the Galaxy we find groups containing four, five or more components. Castor, the fainter member of the two famous Twins, is multiple; there are two main components, each of which is itself a close binary, and at a much greater distance there is a third component which is again a close binary. Castor therefore consists of six stars, four brilliant and two dim red dwarfs – though the colors are not so sharply contrasting as with Zeta Cancri or Zeta Aurigæ.

From multiple stars we come on to loose or open clusters, of which there

The system of Zeta Cancri
Left Diagram of the orbits of the triple star system depicted in the painting opposite. The system rotates about the center of gravity which is offset from the center of the main star. Consequently the orbit of any planets in the system would be subject to a differing gravitational attraction.

Right View from a planet orbiting the triple system of Zeta Cancri. There are three suns; a close pair made up of an orange giant together with a white companion (here looking bluish, by contrast), and a distant yellow star. Gravity distorts the close pair into elliptical shapes, and there is probably an exchange of material from their outer layers. The distances are not represented to an exact scale.

The End of a World

Sometimes a bright star will appear unexpectedly in the sky, remaining conspicuous for a few days, weeks or even months before fading away to its former obscurity. This is known as a nova. The name is rather misleading; 'nova' means 'new,' and a nova is not a new star. It is merely a dim star that has suffered a sudden outburst, flaring up temporarily to many times its usual brilliance.

What would be the effects upon a planet whose sun turned abruptly into a nova? The luckless world would be seared and scorched by the intense heat and the short-wave radiations; any life there would be destroyed, and nothing would be left but a desolate, ruined planet that had been rendered permanently lifeless.

Most stars — including our Sun — shine steadily. A nova is a star in a late stage of its evolution, so that it has used up much of its nuclear energy and has become unstable. Shells of gas are ejected at staggering velocities, and the increase in energy output is colossal. After a while the outburst is spent, and the star returns to something like its old state; but too late to save the planets of its system.

Many novæ seem to be members of two-star or binary systems, but this may not be an invariable rule. There is also a basic difference between an ordinary nova and a 'supernova,' in which the ex-

plosion is so violent that the star destroys itself in its old form, and ends its career as a tiny, immensely dense body made up of neutrons, associated with a cloud of expanding gas. Supernova remnants have been detected, and are generally termed pulsars because they emit rapid pulses at radio wavelengths; one of them lies in the Crab Nebula, which is known to be the remnant of a supernova observed in the year 1054, and which for a time became brilliant enough to be visible in broad daylight.

If a star undergoes a supernova outburst, there can be no hope of survival

for any of its planets. But with a normal nova there may be more hope.

The warning signs

Picture the scene upon a planet which harbors an advanced civilization, and whose sun has started to show signs of instability. The inhabitants — assuming that their scientific knowledge is sufficiently great — will note the danger signals, and will have to decide how best to counter the impending catastrophe. The possibility of preventing a nova outburst is attractive, but we have no idea of how it could be achieved, and it may

well be beyond the power of any technology. There remain two courses of action. The inhabitants can either protect their planet, or migrate.

Protection presumably means 'going underground', but even this will not be practicable if the distance between the planet and the star is insufficient. If our Sun became a nova (a possibility which, let it be added, is so remote that it may be disregarded), the Earth would unquestionably be vaporized. In a similar case, a civilization which was threatened would have no choice but to migrate. This may not be so farfetched as it

sounds, because a nova is an old star; and will have been in existence for long enough to see the emergence of really advanced civilizations. But again circumstances may differ; will there be another planet in the same system which is suitable to receive the refugees?

These are questions that we cannot answer, and yet our remote descendants may have to face precisely this situation. The sun cannot last forever, and in any case it must go through a Red Giant stage, which will menace not only Earthly life, but even the Earth itself.

The dawn of a nova *above*
The sky is still vivid blue, but there is little else to show that the desolate planet once harbored a brilliantly inventive race of beings. The nova outburst has come suddenly, though there has been a period during which the inhabitants of the planet were aware of their danger. From being a mild, benevolent sun, the star has become a deadly destroyer. The temperature has soared to unbelievable heights; the seabed has become molten, and the planet's crust split by severe earthquakes; every living thing has perished. So too have any colonies on the planet's moons, still shing down as they have done for thousands of millions of years in the past. The ruins of great buildings still stand, but they are nothing more than skeletons.

Intergalactic Travel

Will Man ever be able to reach the stars? Today interstellar travel seems as fantastic as flight to the Moon seemed a few decades ago.

The main problem is that of sheer distance. Rockets can fly to the Moon in a few days, but to send the same rockets to even the closest star would mean a journey lasting for hundreds of years. Therefore we must seek other methods. One favorite theme is that of the space ark, in which the original pioneers die long before the end of the journey, and only their descendants make eventual 'planetfall.' Another idea is to put the travellers into suspended animation. Then there are the robots — man-made machines which may pilot space-craft into the deepest reaches of space.

Our first messenger to interstellar space has already been dispatched. Pioneer 10, which bypassed Jupiter in December 1973, has begun a journey which will take it out of the Solar System for ever, and in tens of thousands of years it may bypass another star. It even carries a plaque, so that any alien intelligence that discovers it will be able to identify its planet of origin. But when we consider a journey lasting for this immense span of time, we have to realize that as far as manned travellers are concerned the whole project is impracticable. New methods are needed.

There are two possibilities. One, of course, is to travel as fast or faster than light. Relativity theory teaches that this is impossible, but near the critical velocity one's time scale is altered; this is the celebrated time-dilation effect, which has a solid scientific basis. A crew flying to another star and back, at velocities close to that of light, will complete their journey in a few years by their own clocks; but they will return home to find that centuries have elapsed. This raises all kinds of problems, and in any case it is difficult to see how any velocity of this order could be achieved. Normal

rockets cannot hope to approach it. The best idea is that of an 'interstellar ramjet', which will scoop up the hydrogen spread between the stars and use it as fuel for the motors.

The second possibility is to extend the life span of the crew by some means or other. If the astronauts were 'frozen' soon after the start of the journey, they would remain in a state of hibernation until near their target; but once again an immensely long period would elapse, and the whole project would seem to be of dubious value. An automatic probe, controlled by robots, might be able to

make the journey, but to obtain information from it would seem to be almost impossible once it had passed well beyond the Solar System.

What, then, of the possibility of travel by using some natural phenomenon? Black Holes have come to the fore lately; we do not yet have positive evidence that they exist, but they are at least theoretical possibilities, and there is some observational confirmation as well. A Black Hole is the result of the gravitational collapse of a very massive star. When the star has become sufficiently small and dense, its escape velocity

exceeds the speed of light, so that nothing can escape from it; but there have been suggestions that a spacecraft entering a Black Hole might emerge unscathed elsewhere in the universe.

Finally, there are the theories involving travel by non-material means. Teleportation — the transfer of solid matter through other matter — may seem absurd; but it is no more absurd than sending pictures through the air would have seemed a few centuries ago. This would at once dispense with all mechanical devices, and would also make interstellar travel instantaneous.

A forced landing *above*
We are looking at a scene far away from the Sun and the Earth: another system, another world. Far away we see one of those extraordinary phenomena which seem to be pure fantasy, and which yet almost certainly exist: a Black Hole combined with a giant star. The Black Hole, with its immensely strong gravity, is drawing material away from its companion in a wild spiral. The spacecraft has made a forced landing on a planet in another system, but the environment is bitterly hostile; the area is saturated with deadly X-radiation, and even when heavily shielded no astronaut can do more than venture very briefly into the open. Repairs to the damaged craft can be accomplished only by robots, impervious to radiation.

55

The Photon Star-ship

If it ever becomes possible to travel at near the speed of light, there are endless targets of immense fascination. Planets of all kinds will come within range, the photon rocket and the interstellar ramjet may be capable of conquering these vast distances. Space is not empty, and there is a virtually limitless supply of hydrogen, for use as propellant.

In some regions the interstellar material is sufficiently dense to be excited to luminosity by the radiation from stars in or near it, producing the wonderful nebulae that we can see even today with our telescopes. One of these is the Trifid Nebula, with its shining gas clouds and its dark rifts. It may well contain planetary systems, and in the future it could become one of the targets for photon star-ships.

From Earth, the Trifid Nebula – known officially as Messier 20, or NGC 6514 – is a faint glow in the constellation of Sagittarius. Photographs with large telescopes and long exposures are needed to show it in its full splendor. It is 30 light-years in diameter, and its distance from us is 2,300 light-years, so that we see it today as it used to be before the time of Julius Cæsar. It is gloriously colored, and these colors can be recorded photographically, though they are too fugitive to be seen with the naked eye no matter what telescope is used. Its apparent diameter is almost equal to that of the full moon. In it are bright gaseous regions, excited by the hot stars within so that they shine by their own light; there are dark rifts, where we see the unlit material. Also we can see circular, small dark spots, known as globules and usually associated with the name of a great contemporary American astronomer, Bart J. Bok. It is thought that these globules will eventually shrink and heat up inside, so that they will turn into stars. The Trifid Nebula, then, is a stellar birthplace. Can men ever go there?

There seems no valid reason why not – assuming that interstellar travel will

eventually become possible. The nebular material is very tenuous – many millions of times less dense than our atmosphere, so that a rocket probe could pass through it as easily as through what we usually but erroneously call 'empty space'. But the best view of the Trifid would be obtained from outside it, and the painting shows the scene from a hypothetical planetary system only 60 light-years from its boundary. The planet itself is seen together with a satellite, with the incredible, glowing gas clouds acting as a backdrop. The planet is of terrestrial type, while the rocky satellite more

closely resembles our Moon. The design of a star-ship, such as the one shown here, is of course conjectured: this is assumed to be a 'photon rocket', the giant parabolic reflector emitting a beam of light particles. The light of another 'moon' behind us is reflected by the dark portions of the two worlds. The star around which the planet moves is out of the picture, to the right.

An astronomer on a planet in this position would have a fascinating view of the way in which stars are born; the process is admittedly a slow one, but when a star blows away its surrounding

ring of dust and begins to shine brightly it can do so quite suddenly. No doubt this is happening all the time to the embryo stars in the Trifid. Also, the structure of the gas and dust would change over relatively short periods; there could be no thought of an unchanging sky, such as that in which our ancestors used to believe.

The Trifid is not the brightest emission nebula visible from Earth; the nebula in the Sword of Orion is more conspicuous, mainly because it is closer (1,500 light-years) – but for sheer beauty the Trifid remains unsurpassed.

Journey to the Trifid Nebula *above*
A view of a planet, with its satellite 60 light-years from the huge emission nebula nicknamed the Trifid. The planet, shown to the right, has a terrestrial-type atmosphere, but since its axis of rotation is at almost right angles to the plane of the orbit round its sun(out of the picture, to the right) the cloud patterns are much more symmetrical than ours, and form parallel bands. The large satellite, higher up, is rocky; the light of another satellite coming from 'behind' causes a faint glow on the dark hemispheres of the two worlds. Its motors barely cooling, a star-ship goes into orbit in the life-zone of the planetary system. At its tip, beyond the fuel tanks, shielding and crew-spheres, are small ferry rockets for descent to the planet.

The Asteroid Ark

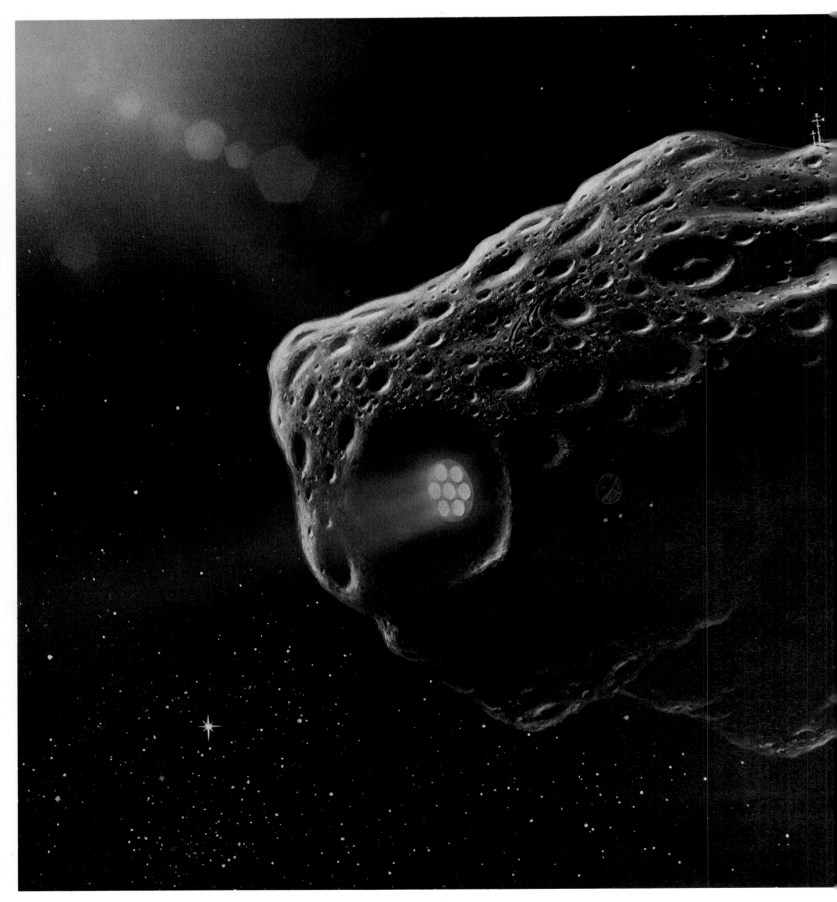

The worst of the many difficulties about interstellar travel is the sheer distance involved. At the velocities below the speed of light that will be used, the time of travel will far exceed that of a human lifetime — or even many lifetimes. One suggestion is to build a 'space ark.' The members of the original crew would be resigned to ending their lives on board the vehicle; only their descendants would reach the intended destination.

Various designs have been proposed; the original idea for a 'space-ark' belonged to J. D. Bernal as long ago as 1929, but it was only in 1952 that L. R. Shepherd suggested that rather than build a vehicle out of metal or some such substance, it might be possible to utilize a natural asteroid that had been hollowed out.

Asteroid probes are already being considered — and there are plenty of available asteroids in the Solar System, ranging from Ceres, over 700 miles in diameter, down to tiny worlds which are not even approximately regular in shape. Some of them must presumably be made up of metals such as nickel iron, and mining operations in the foreseeable future are likely. A small or moderate-sized asteroid could indeed be hollowed out, and it was this which led Shepherd on to the concept of the 'space ark.'

Propulsion would have to be by means of an ion drive, using a low boil-ing-point metal such as cadmium or cæsium. This would yield very high exhaust velocities over a long period, with the expenditure of a relatively small amount of 'fuel,' though even so the fuel tanks needed would be very large. The initial thrust would be small, and high velocities would be built up gradually. Using this method to reach even the nearest star would take many decades, and to reach stars that seem likely to have planetary systems would take longer still, so that the original pioneers would have no chance of surviving to see the end of the journey. Once they

ad left Earth, they would never again know a home other than their asteroid. And if the journey took centuries (as it well might), there would be people who would never know any other world.

Obviously the ark would contain a closed-cycle ecology, with crops and probably even animals being raised for food (though it is quite possible that by the time an ark becomes practicable animals will no longer be used for this purpose). This means that the asteroid would have to be of considerable size. To attempt the experiment upon a small scale would be to court disaster; the

ark would have to contain a hundred people at least, and probably many more. The larger the ark population, the smaller the chance of friction developing between groups of members. It is only too clear that any serious disputes could have tragic results.

For much of the voyage, the ark would be out of touch with any civilization. Contact with Earth would be maintained for a while, but memories and affections would inevitably fade, and second- and third-generation crew members would not know them at all. There would, of course, be extensive libraries and films

on board the ark, but these would be at best a very poor substitute for real experience. One major goal would always be the first contact with another race, and after a while this might almost assume the role of a message from a 'promised land', which might or might not come up to expectations. Once contact were achieved, the whole population of the ark would be forced to make a dramatic mental readjustment. Quite probably the attitude of the crew would by then be very different from that of the pioneers who left their home planet to live in the tiny artificial colony.

A reception committee *above*
The long journey through space is almost over. The asteroidal 'space ark' has covered millions of millions of miles, and is approaching a planet which seems at first sight to be very like the Earth. Deceleration has almost ceased; the ion rockets emit their characteristic glow as they slow the asteroid down preparatory to placing it in a stable orbit around this unknown world.

The people of the asteroid have known no other home. They were born there, and can only have heard of the 'Earth' that their forebears left forever. Now they are ready to meet strangers, and the rockets visible below them show that the planet has an advanced civilization. Whether or not the reception will be friendly or hostile is as yet unclear.

Interstellar *Warfare*

Conflict in space from H. G. Wells's *The War of the Worlds* to the films of the 1970s, notably *Star Wars*, is a classic science fiction theme. Wells envisaged conflict on Earth; over half a century after Wells we can now speculate about conflict between interstellar space colonies. Since Skylab and Salyut, space stations are fact. It is only one step more to the setting up of self-sufficient space-colonies. The stations can be entirely artificial, and therefore different in nature from the asteroidal 'space arks' designed to travel between the stars.

However, the implications of a man-made creation are clearly emphasized in the *Star Wars* story – a space-colony not only designed to travel between the stars, but also to obliterate whole planets.

Man does not have a peaceful nature. The history of the Earth since the dawn of civilization has shown this only too clearly; it is seldom that a decade has passed by without a major war of some kind or other, and there is no reason to suppose that other civilizations in our Galaxy have been more enlightened.

So far as the Solar System is concerned, there is no need to make a space colony invulnerable to attack by aliens, for the simple reason that there are no other intelligent beings. The only danger would come from Earth itself, and by the time that we have become sufficiently

advanced technologically to set up full-scale interplanetary colonies there is at least a chance that the world will have become united. But when interstellar journeys are planned, the situation may well be different.

The asteroidal space ark is above all a vehicle of peace. It is not armed; it has no deliberate defenses, although the risk of being attacked cannot be ruled out (however, this would scarcely affect the crew-members who set out on the journey; they will be dead long before the ark comes within range of any other civilization). On the other hand, it may

be thought that defense is essential, and this involves a completely man-made vehicle, probably spherical in form and with a hull strong enough to make it immune to all conceivable onslaughts. The natural material of the asteroidal ark will be replaced by the toughest materials that Man can devise.

This may in itself increase the dangers. An obviously armored craft appearing from deep space is more likely to arouse suspicion than a converted asteroid, and it will inevitably have a somewhat sinister appearance. Moreover, it will have the ability to 'fight back', possibly

by devices such as lasers — the modern equivalent of H. G. Wells's heat ray. If the reception is hostile, then a full-scale battle will be inevitable, and the travellers, with no reserves or place of refuge, must be at a disadvantage.

Much depends upon the state of evolution of the alien race first contacted. There is a chance that every civilization goes through a 'testing time', when it has the ability to destroy itself but has yet to learn the art of self-control; Earth is at present in this state, and has been so ever since the first atom bomb exploded in 1945. If the wandering colony en-

counters a culture of this type, the danger may be acute. If, on the other hand, the initial contact is with a race that has survived its 'testing time' and has developed a culture that is advanced morally as well as technologically, the situation will be different again.

No doubt both kinds of civilizations exist. If preliminary contacts are established while the colony is still out of range, avoiding action could be taken, but communication problems will inevitably be extremely difficult, and the travellers might not be aware of their danger until too late.

Battle stations *above*
The space colony has arrived near a planet that harbors an alien civilization. It has used the gravitational pull of a neutron star to swing it into its chosen orbit; the neutron star, the remnant of a supernova outburst, is still visible to the lower left. But this time the reception is far from welcoming. The inhabitants of the alien planet regard the visitor from space as hostile, and they have set out to destroy it. Huge rocket vehicles crowd into the attack, but the colonists have perfected their defenses; one of the attackers has already been destroyed by an intense laser beam. It would seem that space has become yet another battleground, that Man's warlike nature is still unappeased. Like so many other fictional concepts, the 'interstellar war' has become hard fact.

Conclusion: The End?

In this book we have looked into the future. Starting with our home, the Earth, we have passed from space-stations to the Moon, the Sun's family of planets, and thence to the distant stars. Yet are we being over-speculative and unrealistic? As our rockets fly farther from our home, it is an appropriate time to take stock of our situation and see what lies in store for us.

There is no need to do more than repeat that only a few decades ago the idea of travel to the Moon was ridiculed. Today transmitting stations have been set up there, and Earth-made vehicles stand on the surfaces of Venus and Mars. Interplanetary flight cannot lie very far ahead. But when we consider the stars, the problems are not hundreds but thousands or even millions of times more difficult. The distances involved are so immense that it is impossible to conceive of them in normally recognizable terms.

Are we justified in saying that the stars are forever beyond our reach? The answer can only be 'no', even though as yet we have no hard and fast ideas of how the journey can be accomplished. To us, a journey to Sirius, to Zeta Aurigæ or to the star-studded core of a globular cluster seems less fantastic than Mariner 9, or even Sputnik I, would have seemed to Sir Isaac Newton.

The illustration on the right imagines a scene at night on a planet of a star in such a cluster. Globular clusters are relatively compact systems. They belong to our galaxy, but all are very remote — over 20,000 light-years away. They form a kind of 'outer framework' to the Galaxy, and are made up of Population II stars, so that their most luminous stars are old red giants that have evolved away from the Main Sequence. They are symmetrical, and may contain upward of half a million stars equal in luminosity to the Sun, together with an uncertain but very large number of dimmer stars which we cannot see individually.

It is not easy to believe that any material vehicle could ever take an Earth-traveler as far as a globular cluster; remember that even moving at the velocity of light, a probe would take over twenty thousand years! Yet there is no reason why the stars in a cluster should not have planet families.

Planet in a globular cluster

The planet shown here is assumed to have an atmosphere sufficiently thick to support life. There can be no night as we know it; the sky is filled with thousands of stars, many of which would be as brilliant as Sirius appears to us, while others would be as bright as our full Moon. The light background to the nearer stars is caused not by atmosphere, but by cloud upon cloud of more remote stars. The prevailing color is warm and rosy, since, as we have noted, the most luminous cluster-stars are red giants. The nearest stars will be only light-months away, but even so the separations amount to thousands of millions of miles, so that stellar collisions must be extremely rare. In the picture, we see a satellite of our hypothetical planet.

The survival of civilizations

The most exciting prospect of all is the chance of contacting other intelligent races. It is logical to suppose that where life can arise, life will arise — and will evolve to the highest degree possible according to its environment. The picture in our scene is quite outside our normal experience — unless perhaps one is reminded of protoplasmic life-forms in the sea — but it is possible to build up a whole 'ecology' for a hypothetical planet. Here the artist postulated massive colonies of oxygen-filled bladders, anchored to land or forming great rafts on an ocean, maturing and finally floating free into a carbon dioxide atmosphere, where they rise and travel in atmospheric winds. Such a scene is of course pure speculation, but one of the biggest questions remaining is, 'Can life take such totally alien forms, or will it, if encountered, be basically Earthlike?'

How many civilizations will survive is another matter. Some, no doubt, will destroy themselves by warfare as soon as they have split the atom and learned enough to render their homelands un-inhabitable. Others will be wiped out by natural calamities. The rest may, we are entitled to hope, progress far beyond our own present primitive state. Their needs and their ambitions may be totally different from ours, but there are certain common factors. Astronomy must be as fascinating to them as it is to us, and at this moment there may be many optical telescopes and radio telescopes pointed skyward — some, unquestionably, in the direction of the Earth.

Whether we shall be the first in the Galaxy to achieve interstellar travel remains to be seen, and it is not a question which can be answered in the lifetime of anyone living in the twentieth century. But so long as life on Earth survives and progresses, so Man will continue his attempts to reach out into the Universe. Apollos and Mariners will give way to deep-space probes, then to star-ships and finally, perhaps, to vehicles which can travel across the vast gulfs between the galaxies. They may not be material ships, but in the 6,000 million years remaining to us before the Sun turns into a red giant we must make our journeys beyond the Solar System. We can never ignore the supreme challenge of the stars.

Alien life forms *above*
The final painting is the most speculative of all, for it attempts to show a totally alien life-form: oxygen-filled sacs which are capable of floating freely in the winds of a carbon dioxide atmosphere. To any inhabitants with senses resembling our own, the 'night' on a planet of a star in a globular cluster would be beautiful and fascinating, with its thousands of stars — many as bright as Sirius, and others as brilliant as our full Moon.